解码智能时代 2021
前沿趋势 10 人谈

信风智库　编

访谈：曹一方　骆啸宇

摄制：龚　洋　刘嘉陵

文稿整理：曹一方　王思宇　胡　潇　欧阳成

翻译：王　蓉　胡文江　李珂雯　魏江渝
　　　刘　悦　张世影　吴　链

重庆大学出版社

图书在版编目（CIP）数据

解码智能时代.2021.前沿趋势10人谈 = Decrypting the Intelligent Era 2021: Interviews with 10 People on Future Technology Trends：英文 / 信风智库编；王蓉，胡文江译. -- 重庆：重庆大学出版社，2021.8
ISBN 978-7-5689-2883-0

Ⅰ.①解… Ⅱ.①信…②王…③胡… Ⅲ.①人工智能—英文 Ⅳ.① TP18

中国版本图书馆 CIP 数据核字（2021）第 139055 号

本书的视频拍摄工作由重庆市九龙坡区融媒体中心、重庆大数据人工智能创新中心、公共大数据安全技术重庆市重点实验室支持完成。

Decrypting the Intelligent Era 2021: Interviews with 10 People on Future Technology Trends
解码智能时代 2021：前沿趋势 10 人谈
JIEMA ZHINENG SHIDAI 2021: QIANYAN QUSHI 10 REN TAN

信风智库 编
王 蓉 胡文江 译

策划编辑：雷少波 杨粮菊 杨 琪
责任编辑：苟荟羽 杨 琪　版式设计：许 璐
责任校对：王 倩　　　　　责任印制：张 策

*

重庆大学出版社出版发行
出版人：饶帮华
社址：重庆市沙坪坝区大学城西路 21 号
邮编：401331
电话：（023）88617190　88617185（中小学）
传真：（023）88617186　88617166
网址：http://www.cqup.com.cn
邮箱：fxk@cqup.com.cn（营销中心）
全国新华书店经销
重庆俊蒲印务有限公司印刷

*

开本：720mm×960mm　1/16　印张：31.25　字数：443 千
2021 年 8 月第 1 版　2021 年 8 月第 1 次印刷
ISBN 978-7-5689-2883-0　定价：88.00 元

本书如有印刷、装订等质量问题，本社负责调换
版权所有，请勿擅自翻印和用本书
制作各类出版物及配套用书，违者必究

前言

向时代提问，让时代回答

从石器时代到农耕文明，从工业革命到互联网时代，人类历史通常都是以不断进化的生产力来命名一个时代。如今，人类正在进入一个"人机物"三元融合的万物智能互联时代。

以新一代信息技术为引领的科技创新，展现出了人类历史上前所未有的复杂性和重要性。大数据、人工智能、物联网、区块链与量子科技等新兴技术，不仅日新月异地更新迭代，而且互相交叉、融合、叠加与裂变，由此引发人类社会、经济、文明的深刻变革与进步。

这种变革与进步，究竟是什么模样？关于未来的各种预言流传甚广，有的毫无悬念，有的不可思议，有的盲目乐观，有的危言耸听。那么，此时此刻我们该如何获得一种既有宏观又有微观，既有思考又有实践，既有现实又有推论的前瞻视角与认知方式？

陶行知说过，发明千千万，起点是一问。提问或许是认识这个复杂且重要的智能时代最好的方式。这就是我们策划这本《解码智能时代 2021：前沿趋势 10 人谈》的初衷：向时代提问，让时代回答。

实现这一想法并非易事——

首先是话题丰富性。我们选取了从大数据、人工智能和区块链，到工业互联网与数字孪生，再到量子科技、深空探测和脑机接口等 10 个热点话题，贯穿了智能时代从基础理论研究，到前沿科技探索，再到技术产业应用的三个维度。比如，大数据专家周涛教授为我们讲解，如何打造新一代数据治理体系；科大讯飞吴晓如总裁和我们分享，人工智能如何给教育和医疗等民生领域带来改善；华南理工李远清教授带领我们探索，大脑如何与"计算机"实现"心灵感应"；重庆大学谢更新教授又把我们的视线带上月球，看看嫦娥系列月球探测器如何一步步实现"绕落回"。

其次是对象多样性。从跨国科技巨头、本土业界龙头、上市公司到创业新锐，我们访问了图灵奖获得者、中国科学院院士、学术带头人和产业领军人，力求从多个角度展现一个更加立体与多元的智能时代。比如图灵奖获得者约瑟夫·希法基思为我们解释，人工智能如何解决不同维度的复杂性；中国科学院徐宗本院士为我们阐述，数理基础研究如何支撑人工智能更进一步发展；法国达索系统集团全球副总裁罗熙文为我们打开了数字孪生的虚拟世界，领略如何无数次地"彩排"我们现实世界的生产与生活；上海交通大学教授、图灵量子创始人金贤敏则把我们带进了量子科技的微观世界，探究神奇的量子如何在计算、通信与测量等领域带来不可思议的质变与飞跃。

再次是问题关联性。树根互联 CEO 贺东东分析了"工业互联网为制造业带来数字化、网络化与智能化"，这正好契合了徐宗本院士对智能时代本质特征的深入剖析，以及达索系统罗熙文副总裁对数字孪生"让虚拟世界和现实世界互为映射"的总结。万向区块链肖风董事长提及"区块链支撑万物互联带来的海量数据"，恰巧周涛教授也谈到了"区块链不可篡改的属性让数据更可信"，而这也被应用在了树根互联的工业互联网平台。无独有偶，肖风董事长、约瑟夫·希法基思教授以及罗熙文副总裁都阐释了"如何管理万物互联复杂性"的重要问题。这种不谋而合与互相印证还有很多。这并非我们刻意设计，而是这些前沿趋势自然而然地汇聚交融，如同百川奔流入海，形成了一股推动人类文明能级提升的巨大力量。

最后是时间延续性。风物长宜放眼量，如果把 2021 年作为一个起点，我们希望这种提问与回答能够延续下去，每年选择 10 个话题邀请 10 位国内外的专家与企业家，见证智能时代未来 10 年的变革历程。技术如何迭代，产业如何变迁，思考如何更新，趋势如何兴起，放在一个年代的时间长度上，这样的提问与回答就具有了一定程度的历史价值。时代由伟人推动，历史由人民创造。向智能时代提问，就是向智能时代的思考者、研究者、探索者与实践者提问。无论哪种角色，都是这个伟大时代不可或缺的组成部分，都是这个伟大时代添砖加瓦的创造者。从这个意义上讲，他们的回答也就是时代的回答。所以，我们向更多的国内外专家、企业家以及创业者发出未来 10 年的诚挚邀请。

当然，向时代提问，让时代回答，都是基于科学规律和客观事实，而不是凭空想象与牵强附会。但无论提问还是回答，都难以摆脱时间与视野的局限性，尤其是在突飞猛进的智能时代，更是凡事无绝对。所以，问与答都绝非权威，而是诚意满满地希望引起大家的关注、思考与探讨。牛顿说，物体之间存在万有引力；爱因斯坦说，不存在引力，是时空弯曲。两位巨匠都闪耀于人类科学的星空。

  向时代提问，让时代回答，终究是有意义的。正如剧作家萧伯纳所言：你看见一样东西就会问"为什么"，而我会梦想那些还未出现的东西，并且问"为什么不"。

信风智库 曹一方

2021 年 7 月

目录

徐宗本：
数学如何让人工智能更智能

数学与人工智能相辅相成　004

突破数理基础，才能让人工智能学会学习　009

数学如何赋能智能化产业创新应用　015

约瑟夫·希法基思：
人工智能如何获得社会信任

解开复杂性的迷思　026

如何建立人工智能的可信标准　030

从万物互联，看智能系统如何升级　034

创新生态系统的三大要素　039

贺东东：
工业互联网如何重塑未来制造

为什么说工业互联网是新型基础设施　044

数字化与智能化之路该怎么走　049

工业互联网将给产业带来哪些改变　057

探索智能制造的中国逻辑　062

周涛：
如何打造新一代数据治理体系

如何激活大量沉睡的数据资源　070

大数据需要"三高"：高标准、高质量与高价值　075

数据要素市场化流通如何落地　079

大数据如何赋能社会治理　083

如何应对万物互联带来的数据大爆炸　087

吴晓如：
智能化如何为生活添彩

人工智能如何开启下一步创新　094

人工智能探索与应用有哪些路径　098

智能化为各行各业带来了哪些改善　102

人工智能如何肩负起社会责任　108

肖风：
区块链如何构建新型生产关系

新型生产关系新在哪里 114

万物互联需要去中心化 118

通证经济将如何影响我们生产生活 121

区块链天然匹配数据要素市场化流通 126

用数字货币服务去中心化世界 129

罗熙文：
数字空间如何与现实世界共舞

数字孪生降低试错成本 136

连接实体经济与数字经济的桥梁 139

数字孪生背后是数据融通 141

数字孪生使城市更美好 145

数字孪生的中国发展与未来趋势 148

李远清：

脑机智能如何孕育"最强大脑"

从"理解脑"到"保护脑"与"创造脑"　154

人工智能如何学习人脑　158

怎样连接人脑和计算机　160

脑机接口的产业化风口在哪里　164

脑机接口还有哪些想象空间　169

谢更新：

月球上的第一片绿叶意味着什么

揭开月球背面的神秘面纱　176

月球上的第一片绿叶　178

仰望星空的民族才有未来　182

金贤敏：
量子科技如何打开人类新纪元

量子科技带来的改变　196

量子科技的产业化之路　203

量子计算的优越性　209

后记　智能时代的年度印记　215

徐宗本：
数学如何让人工智能更智能

徐宗本 中国科学院院士，数学家、信号与信息处理专家，琶洲实验室主任，中国科学院信息技术科学部副主任，西安交通大学数学与统计学院教授、博士生导师，陕西国家应用数学中心主任、大数据算法与分析技术国家工程实验室主任，国家大数据专家咨询委员会委员，国家新一代人工智能开放创新平台及战略咨询委员会委员。

主要从事智能信息处理、机器学习、数据建模基础理论研究。长期从事 Banach 空间几何理论与智能信息处理的数学基础方面的教学与研究工作。提出了稀疏信息处理的 $L_{(1/2)}$ 正则化理论，为稀疏微波成像提供了重要基础。发现并证明机器学习的"徐－罗奇"定理，解决了神经网络与模拟演化计算中的一些困难问题，为非欧氏框架下机器学习与非线性分析提供了普遍的数量推演准则。提出了基于视觉认知的数据建模新原理与新方法，形成了聚类分析、判别分析、隐变量分析等系列数据挖掘核心算法，并广泛应用于科学与工程领域。

曾获国家自然科学二等奖、国家科技进步二等奖、陕西省最高科技奖；国际信息技术与量化管理（IAITQM）理查德·普莱斯 (Richard Price) 数据科学奖；中国陈嘉庚信息技术科学奖、CSIAM 苏步青应用数学奖，并曾在 2010 年世界数学家大会上做 45 分钟特邀报告。

扫码观看
访谈精选视频

在智能时代,信息空间如何与人类社会和物理世界互相作用?我们又该如何理解数字化、网络化与智能化?机器学习乃至人工智能进一步发展,还需要解决哪些问题?人工智能如何掌握学习方法论?中国在大数据与人工智能方面的研究与应用,应当注意哪些问题?

数学与人工智能相辅相成

信风智库:人类社会、物理世界与信息空间构成了如今的三元世界。在智能时代下,信息空间如何与人类社会和物理世界互相作用?我们又该如何理解数字化、网络化与智能化?

徐宗本:这需要我们对世界有一个宏观的认知。人类社会、物理世界、虚拟空间(信息空间)构成了当今世界的三元空间。人类是群居性动物,个体之间会形成相互关系,人类社会就是由这样一个巨大的"关系网络"构建的。而人类生活的地方和空间,则是物理世界。当人类社会与物理世界合在一起,并发生作用,就形成了现实世界。在现实世界中,每一个事物都是有物理载体的。

为什么说人类是具有智慧的生物？答案就是，人类是希望不断改变和完善现实世界的，并愿意为此付出大量的创造力与生产力。比如人们需要正衣冠，所以镜子被发明出来，将现实世界中的样貌投射到镜子中，从而辅助人们精准地整理仪容。

这时，一个与现实世界对应的虚拟世界便产生了。虽然镜子里的事物没有物理属性，但意义和价值在于，我们能通过它认识现实世界。将这种逻辑带到今天，就是我们所说的智能时代与数字经济。镜子里的像可以理解为数据，镜中世界就是虚拟空间，我们可以通过信息空间来折射现实世界，人类社会便会得到快速的发展。

数字化、网络化与智能化是信息空间最显著的三个特点。

首先，感知人类社会和物理世界的基本方式是数字化。数字化将现实进行虚拟化的折射或投影，在智能时代下，通过对现实世界的一个个数据碎片进行拼凑与梳理，便可以认识现实世界的本质与逻辑。数字经济就是利用数字化的方式来呈现并驱动经济发展的。

其次，连接人类社会与物理世界（通过信息空间）的基本方式是网络化。如果要通过碎片化的数据找回真实的现实世界，那么信息之间就要存在某种特定的网络，我们需要找到它并建立一个通道，让信息空间与现实世界相通。这就是网络化。

最后，虚拟空间作用于物理世界与人类社会的方式是智能化。人工智能就是希望能够将现实行为放在虚拟空间中完成，并且达到跟在现实空间完成一样的水平和难度。

用虚拟空间的办法来增强人们认知和改善现实世界的时代，就是智能时代。中国的发展目标提到了"数字中国""网络强国"和"智慧社会"，其内在核心就是要大力推动数字化、网络化与智能化的发展。

信风智库：数学是人工智能未来发展的基础。这体现在哪些方面？同时，人工智能又是如何反作用于数学研究并促进其发展的？

徐宗本：虚拟空间是对现实世界的映射。但是现实世界这么大，怎么才能全部映射出来呢？显然通过"一块镜子"是不够的，而是要通过海量的数据拼凑起一个个片段，从而还原现实世界。这就需要人们在虚拟空间内找到数据的结构运算和演化规律——数学。

实际上，早期的数学仅仅是因丈量事物的需要而产生的一种认知语言，用来描述客观事物。如今数字经济时代，数学本身已经成为一种工具与技术。数学反映现实世界，它是源于现实世界却高于现实世界的一种抽象的存在。从这个角度来讲，数学与人工智能在处理问题的方法论上存在着本质上的一致性，它们之间也有着相辅相成的关系。数学作为人工智能发展的基石，不仅为人工智能提供新的模型、算法和正确性依据，也为人工智能发展的可能性提供支持平台。这主要体现在三个方面：

第一，数学为所有人工智能技术的演化提供了方法论，没有形式化就没有计算机化，而数据是形式化的基础；第二，数学提供了模型，任何技术研究，如果不能写成数学模型，就不可能深入，就不可以从感性试验走到理性模型；第三，数学还为人工智能提供了直接的语言和工具。

人工智能是模拟人的行为或者能力，即在既定环境中，智能体可以通过与环境的交互来适应环境，从环境中获取信息并进行自我思考，从而提高自身解决问题的能力。这跟数学一致，都是通过获取数据、理解数据，反过来提高认知、解决问题。

比如，从数学的维度来看，机器学习其实就是一个函数空间或参数空间的优化问题，两者在本质上并行不悖。

在人工智能的作用过程中，非常重要的一个部分就是环境。一方面，可以通过数据、信息来描述环境；另一方面，可以用数学的方式对环境进行建模，通过智能体在环境模型中的行为判断，修正智能体行为的指标。

换句话说，环境通过数学模型生成了一种规则，而智能体在这个规则下能够实现最优的决策。以人为例，人在天热时脱下外套，虽然脱下外套是一个下意识的动作，但其实也是人脑在规律模型下经过运算分析后做出的最终决策——因为脱下外套是最有效率的散热手段。

数学不仅为人工智能创造了新的规则模型，也提供了算法和正确性依据。反过来，人工智能深度学习算法也可以很好地反哺数学研究。比如超参数设定、复杂的偏微分方程求解等数学问题，如今都可以用人工智能的算法模型和大数据分析来解决。

举一个简单的例子，天气预报以前主要通过自然周期变化来推算天气的走势。而现在假如一个地区要召开奥运会，需要提前了解到这个地区 24 小时的精准天气情况，传统的方式就无法实现了，这就需要精准天气预报。

那么精准的天气预报如何实现呢？其本质就是一个复杂的气象要素偏微分方程组问题。它通过在全球各地建立气象站台，监测实时的气象要素数据，及时共享到地方气象数据后台，然后利用天气预报数据模型进行要素数据分析，从而解出未来大气的运动状态，最后算出精准的预报结果。

精准天气预报的难点在于它需要大量的运算，因为天气情况是一个蝴蝶效应，微小的误差会导致结果大相径庭，所以这个方程是非常复杂的。在没有计算机的年代，有些研究者花了一年的时间才解出第二天的预报，而且实际情况很可能截然不同。人工智能的出现无疑提供了超强的算力支持，使精准天气预报成为可能。

信风智库：算法最初其实是数学概念，即解决某个问题的计算方法。传统意义上的算法与人工智能的算法有着怎样的关联与异同？

徐宗本：严格来讲，二者在教育术语上并没有差别。算法本质上是一个规范化的逻辑程序，是用来解决某些问题的。细分下来，算法有很多种类，其中，数学算法也称为授权算法，是比较原始的，是基于数学逻辑来做的，所以不考虑它的应用。

还有一类计算机算法，通常是指根据计算机结构特征优化以后得到的数学算法。因为每一个计算机都采用了不同的通信原理和不同的调度方法，这就需要不同的操作系统去实现，所以使用这种算法是为了让计算机的运行更加节约、更加优化。

我们看到市面上很多大数据平台，都强调自己采用了人工智能算法，其实大部分都是经过优化以后的计算机算法。从这个角度上来讲，数学算法、计算机算法和人工智能算法之间没什么本质差别。

人工智能算法则是具有一些显著特征，比如强调处理数据的规模很大，但其本质上还是数学算法，包括很多抽象的微分方程、线性方程组以及传统的解数问题算法等。所有通过人工智能解决问题的算法，最终落脚点都是解决数学问题，我们只需理解这个就够了。

突破数理基础,才能让人工智能学会学习

信风智库:从数理基础的角度看,机器学习乃至人工智能进一步发展,还需要解决哪些问题?

徐宗本:这是个很深刻的问题。很显然,人工智能已经突破了从"不可以用"到"可以用"的技术拐点,正在从"可以用"迈向"很好用",处在从人工化走向自动化的"前夜",迈向自主化的初级阶段。而它的发展也不仅仅是用到了线性代数、统计学和概率论等背景学科知识,人工智能的核心能力应该是算法、算力和算据,即以深度学习为代表的模型、以超级计算为代表的计算资源以及以大数据为代表的数据资源。基于这三种基本能力,人工智能才变成了可以应用的技术。

中国在人工智能应用方面走在世界前列,因为我们有大量的数据资源,这是天然的优势。我们将人工智能应用到各个领域,探索了很多模式,也实现了较好的变现。但是,从应用研究上来看,目前的发展阶段与人工智能的真正目标相比,还相差甚远。人工智能的最终目标是利用人工智能的基本逻辑,搭建一个系统智能体,像人一样去适应环境,实现自由决策、自动反应和自主行动,代替人去做一些繁杂的劳动,切实解决现实社会中劳动成本高的问题;代替人从事一些高风险的工作与

探索，让人们远离危险，真正意义上解放劳动力，提升生产力，而不仅仅是刷脸、线上支付等初级应用。实现这一目标，还存在诸多技术瓶颈，亟须重大技术变革，我们还有很长的路要走。

虽然目前的人工智能也运用到了数据分析，但还不具备像人一样的思考能力。它还需要解决以下几个问题。

首先，是形成新的统计学理论。传统的统计学是基于两个概念，一个是"独立同分布"的基本假设，简单来说就是事物之间是独立的，但它们都具有一定的系统结构分布。基于这样的假设，抽样调查便成了可能。因为抽样里的部分样本，都会服从总体样本的分布。

另一个就是概率论。比如一对夫妇生男孩或生女孩，概率是一样的；又比如掷硬币，正面和反面的概率也是一样的。但是，它们必须有一个前提，就是操作的次数要无穷大，才符合概率论的理论支撑。这也叫作无穷远形态或者大样本性质。

传统统计学更注重数据呈现的结果。但实际上，如果数据量有限，其结果也不足以形成好的决策方案。而大数据与人工智能的发展则改变了这一思维方式，它更看重问题的本质。大数据的核心是从传统统计学理论中找到细微的指标，然后通过人工智能大量采集核心数据并对指标进行运算，从而得出更具指导性的结论。所以，以大数据为核心的新统计学理论不是对传统统计学的颠覆，而是在它的基础之上进行补充和升级。

其次，是建立以大数据计算为基础的新算法。举个很通俗的例子，我们把数据分析应用比喻成做菜，以前数据样本少，油盐等调料和食材都放在锅的旁边，需要什么、需要多少直接拿，不管好不好吃，至少一盘菜很快就做好了。但现在是大数据时代，食材和调料多而杂，只能用仓库来储存，而且还有很多都放错了位置。在这种情况下做菜，便需要用到新的方式，也就是新的算法。

这些算法的本质就是线性方程组求解、图计算、最优化计算以及高维积分等数学问题。比如地图智能导航从西安到北京怎么走，过去地图分辨率不高，根据普通的地图可以获取基本的路线。但随着信息技术的发展，地图的分辨率越来越高，不可能一次就涵盖西安至北京之间全部城市与道路的数据，只能一次次地分别给出其中某些城市之间的道路信息。那么怎样走最近、要带多少钱，其实就是在解决分布式图信息环境下的图计算基础算法问题。

最后，是深度学习的问题。人工智能的核心是深度学习，但是如何解释深度学习后的结果是未来需要解决的问题。比如智慧医疗的应用，人工智能通过对积攒的病例大数据进行深度学习，找出规律，从而实现机器自动诊断。但是，一个病人来检查，机器告诉他得了白血病，他不接受，你如何解释呢？机器深度学习的过程极其复杂，并且通过黑箱运算，没有办法解释。

又比如智慧金融，一个客户来贷款，风控系统对其经济状况、社会关系等数据模型进行运算，得出了不符合贷款要求的结果，客户也不能接受，需要一个解释，你也无法给出解释。

深度算法是人工智能最基本的特征。虽然并不是每一个应用都能解释，但是如果我们想要将人工智能技术应用到更多的领域和场景中去，有些决策也不得不解释。换句话说，人工智能想要得到更高的发展，还需要解决很多基础的数理问题。

徐宗本院士总结的人工智能十大数理基础问题

1. 大数据的统计学基础
2. 大数据计算基础算法
3. 数据空间的结构与特性
4. 深度学习的数学原理
5. 非常规约束下的最优输运问题
6. 学习方法论的建模与函数空间上的学习理论
7. 非凸优化的理论与高效算法
8. 如何突破机器学习的先验假设
9. 如何实现机器学习自动化
10. 知识推理与数据学习融合的 AI 系统建模与分析

信风智库：元学习是学习如何学习，是弱人工智能迈向强人工智能的重要突破口之一。那么，让人工智能掌握学习方法论意味着什么？我们目前已经进行了怎样的探索，还要解决哪些问题？

徐宗本：这是一个很好的问题。现在的人工智能实际上还处于完成任务的学生阶段。怎么理解？我们要做一件事情，就要为人工智能搭建一个系统，这个系统只限于完成这一个任务。如果要执行其他任务，就还需要搭建另一个独立的系统。简单来理解，就是目前的人工智能只会做事，却没有方法论，所以我们称它为弱人工智能。

但很显然，我们对人工智能的长期目标是希望它能够真正地模拟人去适应环境、理解语言，甚至学会独立思考，这个目标就是我们追求的强人工智能，也是老师阶段。强人工智能的特点就是具备了学习方法论的能力，也就是元学习。

那么什么是超人工智能呢？其实这是另外一个维度的概念。首先我们要理解，"超"其实是一种能力的扩展，并不是一种人格智能。技术的本质是延伸人体器官功能的方式方法。人类为了更好地生存，就通过技术创造了比人体本身更强的能力。

比如人跑不快，就发明了汽车；人运算能力有限，就发明了计算机；人为了能够上天，发明了飞机，这些都是人体自身做不到的事情，但是技术为人类实现了。所以从这个维度来理解，超人工智能并不是比人更有智慧，甚至取代了人类，它只是在某些特定能力上比人类更突出，是对人工智能的一种形容。

实际上，弱人工智能、强人工智能和超人工智能是交错发展的。现在很多机器智能已经超过了人类的智能，比如计算效率，计算机不但速度快，且不知疲惫，也不会受情绪影响。为什么阿尔法狗能够打败围棋高手，有一部分原因就是它不知疲倦，也不会被输赢所影响，这是它的优势。

这里我再举个例子，自动驾驶其实就是汽车在行程当中自动完成一个又一个的任务。如果我们把一辆汽车放在一个封闭的空间内，给它设计好特定的程序，它实现全路程自动驾驶当然没有问题。但是，如果放在自然的交通线路上，它就无法完成了。因为设定好的程序只有一个固定任务，那就是按照线路跑完全程。但在实际道路中，会发生很多随机的情况，比如起大风了、风沙来了、有其他车辆变道超车或者有行人突然越过障碍等，这时候任务就变了。

这背后其实就是元学习的问题。我们给汽车装上了摄像头、雷达等各种各样的传感器，其本质是收集各种各样的环境信息，然后对信息进行处理，最终自动做出决策。

以前，信息任务的自动化处理是分离的，比如去噪、跟踪与光平衡等任务，机器无法根据环境变化自动切换任务。而元学习就是让机器具备学习各种任务处理的学习能力，也就是让机器掌握一种应对各种问题的学习方法论。这样一来，机器就能根据环境信息，自动切换各种信息处理任务，自动形成新的解决方案，比如自动驾驶汽车探测到周围的环境变化，及时做出反应，如减速、变向或停车等，甚至选择最优的路线继续行驶。

数学如何赋能智能化产业创新应用

信风智库： 实际上，大数据智能化也为数学提供了更为广阔的应用空间。在您的研究与实践中，具体体现在跟生产生活相关的哪些方面？

徐宗本： 我举一个医疗领域的应用案例，比如 CT 诊断就是典型的用数学建模、用大数据分析诊断的应用。我们知道 CT 诊断的基本原理是通过 X 射线对人体需要检查的部位进行扫描，然后将扫描的光信号转变为数字信号，输入计算机进行处理分析，以此实现对人体内部器官的病变识别。

虽然传统的 CT 诊断应用广泛，但也存在两个最大的问题。首先是它的辐射较大，其实白血病的主要来源之一就是 CT 的辐射，所以医院一般都会要求患者一年只做一次 CT。

其次就是 CT 设备的部署，主要集中在三甲医院。一来三甲医院的影像科人满为患，排号往往需要排两三天；二来农村的患者很难就近享受到 CT 医疗服务。这些都直接导致了老百姓看病难、看病贵的问题。

那么，这两个核心问题如何用大数据人工智能技术解决呢？

第一个层面是体系上的，简单来说，就是把传统 CT 诊断的扫描和成像过程进行分离。我们知道，扫描就是照 X 射线获得信号的过程，而成像是洗 CT 胶片，也就是将光信号转换成数据和图像的过程。如果我们能将这两个板块分离，让两个设备单独运行，就可以解决很多的问题。

我们可以将只负责扫描的 CT 终端机部署到任何地方，包括广大农村地区以及医院里的门诊科室，可以很方便、很及时地对病患进行扫描。而成像的部分则放在中心城市的三甲医院统一集中运用。这样一来，就实现了对医疗资源的进一步合理配置，也符合国家提倡的分级诊疗制度。

那么有人会问，扫描和成像的中间过程怎么解决呢？传统的 CT 设备将二者集中在一个地方，是因为信号传输受限，但随着 5G 的应用，通信传输效率也实现了质的提升，完全可以实现 CT 扫描信号远程同步实时传输，而成像中心的底片数据也可以实时反馈回各个终端，如此就构建了全新的 CT 诊断流程。

第二个层面是技术上的，用数学模型的方式重构 CT 扫描模型，最大限度地减少辐射。传统的 CT 通过高能量粒子构成 X 射线对人体内部进行扫描，而要想清晰成像，就必须匹配相当高的电压和电流，使用大剂量的 X 射线，但强烈的 X 射线在穿透人体时会影响细胞功能和代谢，从而引发临界点的癌细胞变异。这就是辐射的基本原理。

如果通过大数据人工智能，我们就能改善这一情况，核心就是用计算换剂量。怎么理解？我们可以利用大数据建立一个判断图像数据的数学模型，只需用少剂量的 X 射线扫描一部分关键信号，然后将这些信号放入数学模型进行比对，最终也能分析出符合实际的数据结果，进而实现精准成像。

实际上，数学模型重塑了原始投影和投影数据之间的关系。在这一基础上，CT 扫描就可以将 X 射线的剂量降到以前的 1/5 甚至 1/10。

通过这两个层面的结合，我们就打造了一个全新的智能化医疗产品——分布式微剂量 CT，目前已经通过了相关机构的测试，逐步成熟并将量产。这个产品真正意义上把 CT 设备变成了打印机，我们可以将 CT 扫描终端投放到各个门诊，每一个城市只需建立一至两个影像中心，就可与周边的终端设备完成联动。

再分享第二个典型的案例，就是通过大数据智能化的方式提高核磁共振的检查诊断效率。我们知道，核磁共振的基本原理其实就是将人体置于特殊的磁场中，用无线电射频脉冲激发人体内的氢原子核，引起氢原子核共振。在停止射频脉冲后，人体内氢原子核按特定频率反馈信号，继而被体外的接收器收录，后经计算机处理形成图像。

核磁共振需要人在设备内持续与磁场寻求共振，这个过程需要耗费大量时间，短则几十分钟，长则一两个小时，这对于儿童以及精神状态不好的病人来说就特别痛苦，所以对于核磁共振来讲，大数据人工智能要解决的就不是辐射问题了，而是时间问题。

我们通过大数据可以打造一个快速精准的算法模型，大大提升核磁共振的检查效率。简单来理解，每个人的体内结构是不一样的，所以传统的核磁共振磁场需要花很多时间去覆盖人体内部，从而获得反馈信号。而通过算法模型，磁场就可以在人体内快速找到一条最优的路径，用最短的时间获得信号反馈。

怎么理解呢？就像我们用手机拨打电话，都会有"嘟嘟嘟"的导频声音，这就是在寻找最优的通信频段，从而接通信号实现通话。我们就是把这个原理引入到了提高核磁共振效率的研究当中。

我们设计的核磁共振，会先花一点时间采集病人的身体数据，然后快速按照病人的检查需求，通过一些复杂的数学公式，设定好一个最优信号获取与成像的方案，最终就能实现因人而异的核磁共振快速成像。目前，这项技术已经实现，可以将核磁共振的速度提升 20~30 倍。

其实，这两个案例不仅仅是解决了医疗设备的智能化提升问题，我们还能从中看到更大的价值。比如目前智慧医疗最大的瓶颈就在于医疗数据共享，我们可以通过医疗设备内在的大数据智能化应用，更全面且标准更统一地采集医疗大数据，而这些病例与诊断的大数据还可以在未来的智慧医疗中产生更巨大的价值。

信风智库：相比于欧美国家，我们在大数据人工智能方面的研究与应用，还应当注意哪些问题？

徐宗本：就算法本身而言，其实它跟数学一样，一般是公开的，并不保密。在大数据时代，可以说算法即技术，因为很多信息技术的应用都是大数据智能化成果的转化。目前市场上出现的各种智能产品，比如智能机器人与智能家居等，其内在逻辑都是基于人工智能算法，只是外在通过硬件的方式表现了出来，成为各种各样的产品。

当算法与大数据的场景相结合，就可以研发出一个很好的应用产品。在这方面，中国和国外的理念略有不同。实际上，对于数学算法原型本身的研究来说，国内外的水平其实相差不大，因为我们的逻辑推理能力是非常强的，我们能够在学术研究上实现很有价值的突破。但是，在对算法机理的精益求精上，我们重视程度还不够，往往醒得早、起得晚。

这是让我非常痛心和可惜的地方。有时候我们的学术研究过于急于求成而没有沉淀。比如我们提出了一个新的算法原型或者思想，就急于发表论文，展示成果；而国外大公司看到这个思想成果后，就会琢磨这个算法如何更加精细化，有没有可以落地的模式和场景，如何进一步深化研究与应用。对这些国外公司或机构而言，它们更注重技术的进一步研究与深化，从而创造更高的科研价值和市场价值。

这就跟制造业中的某些情况一样。有些制造技术的原型可能是我们自己创造的，而我们发布以后被国外的企业或机构引用，它们在这个基础上继续精益求精地钻研，最终将新理念、新技术与新工艺固化为一个系统或一个产品。而回过头来，我们看到这个系统或产品时，反而看不懂了。

所以，作为科学家，其实我们是非常忧虑的。我们科研和产业发展始终要解决一个问题，就是克服浮躁，要把事情做得更扎实，眼光看得更长远。从这个角度来讲，商业上某些挣快钱的资本逻辑，实际上就和科研逻辑有冲突。资本要求的是快速孵化、快速迭代与快速换项目；而科研真正需要的是从一而终，慢工出细活。

信风智库：大数据智能化产业是一个体系化、价值化的智慧生态，那么从国家战略、区域发展以及智能产业的角度来看，我们如何把这个生态构建好？

徐宗本：新一代信息技术的研究与应用，可以分为基础研究、技术研究和应用研究。对大数据产业链而言，可以分为数据采集、数据供给、数据存储、数据加工与数据应用等环节，只有链条上各个环节有机衔接，这条产业链才能真正意义上形成生产力。

那么，目前我们大数据产业的问题究竟出在哪里呢？其实就在于产业链不健壮，各个环节的资源配置还不够合理，有的环节产能过剩，有的环节产能严重不足，所以产业链整体上无法充分地形成生产力。

一个很典型的例子就是，很多地区盲目地建设数据中心，使得数据存储这个环节出现了产能过剩。这里面也有一个误区，认为发展大数据产业就是建设数据中心，或者认为数据中心是很显性的产业价值体现形式。

事实上，数据存储也的确是大数据产业链上非常重要的一环，可以很好地支撑通信运营商、科技型企业与平台型企业发展。因为数据中心的建设、运维和能耗的成本非常高，集中建设与运营的数据中心，则可以享受低价的电费和低价的厂房，这些企业就不需要单独建设数据中心了，大大节省了成本。

但是，从更宏观的全产业链角度看，仅仅是数据存储就能产生 GDP 吗？就好比我们把大米堆满了整个仓库，但不拿来做饭，就没有意义。数据中心一定要为数据运营服务。相比于数据存储，我们在数据供给和数据分析等环节上产能不足，我们更应该重点思考如何完善这些环节，这就又回到了重视基础研究上。

其实，人工智能的核心是算法。如何基于现实存在的问题构建解决问题的算法模型，这才是关键。说实话，算法以及背后数理基础的研究，是看不见摸不着的，其成果也不能显性地表达出来，也就不容易引起大家的注意。然而，这种看不见"软实力"的基础研究，不管对于产业链的健壮发展，还是对于科技自主可控，都可谓至关重要。

可以看到的是，我们国家对这方面越来越重视。随着科技创新体制改革的不断深化，作为科研人员，我们看到了更加广阔的前景。在政府的大力支持下，我们在广州建立了琶洲实验室，聚焦人工智能基础理论与核心算法、人工智能软硬件平台与关键技术，以及数字经济重点行业示范应用等研究。我们科研人员在很大程度上有了自主权，比如人才选拔任用、找项目定项目等方面。这充分激发了我们团队的积极性，使科研人员充分发挥专业所长，形成一套面向产业应用的基础科研创新模式。

与其他基础科研领域不同，大数据与人工智能更需要产学研的融会贯通，因为数据和算法本身就来自日常的生产生活，也是为了解决生产生活中的各种实际问题。那么，我们需要重视的是如何更好地将科研成果转化为产业价值，这就需要以市场为导向、以企业为主体的创新。

现阶段我国的科研创新主要还是由高校与研究所等科研机构来实施，具备独立研发能力的企业少之又少。这里面临的问题是，科研机构做出来的创新成果，很有可能离实际需求太过遥远，也不能实现产业价值转化，使得企业不愿意"买单"。所以，大数据与人工智能的产业创新，一定需要建立以企业为主体的自主研发创新生态。

约瑟夫·希法基思：
人工智能如何获得社会信任

约瑟夫·希法基思（Joseph·Sifakis）

法国工程院院士、法国科学院院士、中国科学院外籍院士、欧洲科学院院士、美国艺术与科学院院士、美国工程院院士、法国国家科学研究中心研究总监、国际著名嵌入式系统研究中心 Verimag 实验室创始人。

约瑟夫·希法基思是"模型检测"技术的发明者，因将模型检测方法应用于实时系统验证而在国际上享有盛誉。因其在模型检测领域的杰出贡献，他和卡内基梅隆大学爱德蒙·克拉克（Edmund Clarke）、得克萨斯大学奥斯汀分校艾伦·爱默生（Allen Emerson）共同荣获了 2007 年度图灵奖。

约瑟夫·希法基思的主要研究领域为模型检测及嵌入式系统设计与验证，他提出了通过对时序逻辑公式的计值来验证并发系统性质的思想，包括对含有"可能"和"必然"模态算子的分支时序逻辑的不动点刻划。此外，他还提出了带有"until"算子的分支时序逻辑来表达"公平性"。其模型检测已被应用于计算机硬件、软件、通信协议、安全认证协议等领域，成为分析、验证并发系统性质的重要技术，被英特尔、IBM 和微软等公司广泛应用。

扫码观看
访谈精选视频

我们应该如何理解复杂性？智能化系统要达到可信赖的标准，需要具备哪些条件，解决哪些问题？弱人工智能迈向强人工智能还需要解决哪些问题？从技术与规则两方面来看，自动驾驶还需要解决哪些问题？万物互联将带给我们哪些挑战，我们又该如何应对？

解开复杂性的迷思

信风智库：从本质上讲，智能化是为了解决复杂性的问题。我们应该如何理解复杂性？

约瑟夫·希法基思：复杂性是指解决某个问题的难度。人类通常面临两种类型的问题：一种是我们试图理解甚至预测世界；另一种是我们试图改变世界，以改善我们的生活。

根据问题的类型和性质，复杂性也分成多个维度。

最常见的是计算复杂性，它研究的是计算机求解问题所需要的计算资源（通常是时间和空间）的量。只要问题的规模固定，我们便可以预估出求解的时间。但当问题呈现出规模增长时，我们所需要的计算资源便会不断暴涨。比如，1个开关有"闭合"与"打开"两种状态，2个开关有4种状态，3个开关便有 8 种状态，那么当开关数量为 3 万甚至 30 万个时，每增加 1 个开关，复杂性就会呈指数级增长。系统设计也是如此，每增加一个关键因素，它的复杂性就会呈指数级增长，计算时间也会不断增加。

除了时间之外，计算空间也是我们需要解决的问题。当面对一些难度较高的计算任务时，即便我们能够编写出对应的算法来进行运算，也很难得出计算结果，因为执行这些算法所需的内存数量可能会装满整个宇宙。

第二种复杂性来自可预测性的局限，它是指系统对一个实验或现象的不确定性。这样的问题在生活中非常普遍，比如我们只能在一定范围内，以一定的概率预测某地的天气情况，或者预测某个电子此刻的运动位置。

智能化系统中的可预测性问题尤为明显，以自动驾驶汽车为例，它们必须罗列出环境变化有可能带来的各种结果，并及时采取措施避免危险发生。

第三种复杂性是指可解释性的缺失，业内称这种类型的复杂性为"认识论（来自希腊语'episteme'，意思是科学）"复杂性。当我们需要描述事物之间的因果关系时，必须使用一个科学且严谨的数学模型，并针对不同情况设置不同的模型。例如，物理学中找不到一套理论能够同时解释微观和宏观世界，所以出现了量子力学等阐述微观世界的物理科学。

深度学习等人工智能算法的出现，起初是为了解决认识论的复杂性问题，但现在看来这些方法仍然存在较大的局限。以辨别猫狗照片为例，深度学习系统和人类都可以辨别出两者的不同，人类依靠的是猫狗在形体、毛色和瞳孔等诸多细节上的差异，但深度学习系统却难以解释清楚如何完成识别。这种"不可解释性"恰恰降低了人工智能的可信度，这也是为什么大部分人工智能系统无法在全球获得广泛认可的原因。

第四种复杂性来源于人类在语言表达方面的局限，我们把这种类型的复杂性称为"语言复杂性"。虽然人类拥有丰富的情感和情绪，却无法用精确的语言逐一表达出来。随之而来的问题是，当我们要建立一个与人类水平相当的智能系统时，无法为其建立相匹配的情感和情绪系统。

第五种是设计复杂性。如何理解设计复杂性呢？比方说我是一个设计师，需要设计一个系统。在这个系统的执行平台上有诸多分散式的单处理器系统，还有移动系统和自组织系统。一个系统在执行层面上的板块越多，运行起来就越复杂，也就越容易出现问题。这要求我们在设计人工智能系统时，要尽可能地对系统进行精简，避免后续运行过程中出现问题。

信风智库：智能化解决复杂性问题，是通过怎样的途径？有哪些特点？

约瑟夫·希法基思：需要强调的是，人类与机器解决问题的方式完全不同。人类有两种思维模式，即缓慢的"有意识思维"以及快速的"无意识思维"。

以常见的走路和说话为例，完成这两个动作需要大脑进行复杂的计算，但具体如何完成这个过程，我们尚未研究清楚。要知道，人类在幼年时就能熟练完成双腿移动和语言表达，这是通过长期外界刺激与训练逐渐培养出来的无意识思维。换句话说，大脑可以直接对无意识思维做出反应，无需"思考"的时间。

相反，另一种类型的思考是缓慢而有意识的，它适用于大脑解决复杂的问题。例如，从一个地方到达另一个地方，解决一个数学难题，做一顿丰盛的晚餐，等等。当我们进行有意识的思考时，会对问题进行逻辑上的分割，明确每个部分需要完成的任务。

正是这种"无意识思维 + 有意识思维"的方式，构成了人类的智慧系统。

为了实现人工智能的终极愿景，科学家们也对计算机进行了同样的改造。我们知道，传统计算机依靠程序员编写的代码进行运算。通过这些代码指令，我们可以知道计算机是遵循怎样的规则完成运算的，便于后期检查和追溯。通过模仿人类大脑神经网络结构设计的深度学习神经网络则不然，虽然它具备完成复杂任务的能力，但人类只能看到计算的结果，却搞不清学习的过程。

智能化水平取决于解决三类不同层级问题的能力。

第一类问题是理解世界，即处理和分析感官信息的能力，准确理解环境中发生的事情。例如，理解一段文字，或分析周围环境中物体的运动情况。

第二类问题是用感官信息更新我们头脑中的知识，并通过推理找到所有可能的结果。比如，我收到一封关于体检的电子邮件通知，它需要我查看日程安排，并将体检的日期和注意事项存储在记忆中。

第三类问题是统筹规划，为了达到最终目的而对实现路径进行分层。以体检为例，在体检当天，我们必须计划如何从家到医院，先检查哪个项目，需要花费多少时间等。

总的来说，智能化的最终愿景在于全方位理解世界，把现有的知识与感知到的信息联系起来，并通过层级化部署，最终实现目标。

如何建立人工智能的可信标准

信风智库： 一个智能化系统要达到可信赖的标准，需要具备哪些条件，解决哪些问题？

约瑟夫·希法基思： 智能化系统可信赖的重要标准，便是能够保证其行为符合预期。换句话说，它们不会做对人或环境有害的事情。当然，这取决于智能化系统在场景中的重要性。

如果一个玩游戏的机器人或一个智能助理做错了什么，可能问题不大。但是，如果一个自动驾驶系统或智慧能源系统出现故障，便会造成无法挽回的损失。核心在于，我们需要保证关键系统能够提供人类可以预期的服务，其故障不会造成任何生命或金钱的损失。

正是基于这样的考量，智能化系统必须按照相关标准进行研发。这些标准规定了研发企业应该提供哪些数据，以保证智能化系统的可信度。

以飞机为例，当一个民用飞机被开发出来时，制造商为了获得民航当局的飞行批准，必须证明飞机故障率每小时低于十亿分之一。如果制造企业达不到这个标准，那么他们所研发的飞机也就无法获得飞行审批。

智能化系统的可靠性，需要通过三类标准进行审查。

第一类标准是系统的基础安全要求，也称之为技术的鲁棒性。技术鲁棒性要求人工智能系统必须采取可靠的预防措施防范风险，即尽量减少意外伤害，并防止不可接受的伤害发生。这一要求也适用于人工智能操作环境发生潜在变化的情况，或其他智能体以攻击方式与系统进行交互的情况。

第二类标准是系统的防御安全要求，即系统可以抵御攻击和恶意行为，这也是我们熟悉的网络安全问题。与所有软件系统一样，人工智能系统也应该设置周密的防御措施，以免不法分子利用漏洞进行攻击，例如黑客攻击等。一旦人工智能系统受到攻击，数据和系统的行为就可能被改变，最终导致系统做出不同的决策，或者导致它完全关闭。除此之外，人工智能系统应该有备用安全措施，以便在出现问题时启用备用计划。

第三类标准是系统的性能要求，即系统在面对用户需求时的表现，如运算能力、吞吐量、带宽、抖动、延迟和质量等。一个性能差的系统，即使满足基础安全和防御安全，也可能毫无用处。

信风智库：我们看到自动驾驶在全球范围内时常出现交通事故。从技术与规则两方面来看，自动驾驶还需要解决哪些问题？

约瑟夫·希法基思：自主运输系统的出现，被业内看作迈向人工智能时代的里程碑。自动驾驶汽车是非常具有代表性的案例，它所面临的问题充分说明了人工智能领域发展受到的阻碍。

尽管科技公司和车企们热情高涨，甚至参与了大量的投资和研发，但关于自动驾驶"指日可待"的预测难免过于乐观。目前来看，由于技术局限和公众信任问题，车企们的雄心遭到了不小的打击。就连特斯拉的 CEO 埃隆·马斯克也承认："如果人工智能无法满足现实社会的标准，那么自动驾驶也就无从谈起。"

在我看来，当前的自动驾驶系统设计不够严谨，社会对它们的普遍态度可以概括为"接受其风险"和"贪婪其利益"的现实主义。车企们普遍认为传统而严谨的设计方式具有局限性，不利于自动驾驶的发展，它们更倾向于用过去的经验来解决未来的问题。部分企业甚至乐观地认为自己已经掌握正确的方向，技术实现只是时间问题。

从技术角度分析，自动驾驶还有许多问题尚未解决，仍然需要很长的时间才能实现真正的无人驾驶。我们正在从小型的、中心化的、自动的和小范围可用的系统，迈向复杂的、去中心化的、自主的、大范围的系统。这需要建立一个新的科学和工程基础，而不是简单组合过去 20 多年的现有成果，只关注诸如自主计算、自适应系统与自主代理这类软件。

虽然无人驾驶暂时无法实现，但高级驾驶辅助系统（Advanced Driving Assistance System）正在不断取得进展，ABS、防撞系统和车道辅助等功能，已经在现有的汽车中成熟运行。

如今，车辆在有限且安全的标准下，已经能够实现部分自动驾驶的功能。比如欧洲许多国家已经着手研发自动驾驶卡车。瑞典自动驾驶初创公司英瑞得（Einride）联合欧洲物流供应巨头德铁信可（DB Schenker）在瑞典对 T-Pod 的商用自动驾驶卡车进行了试验。

一辆 T-Pod 卡车长 7 米，能运载 20 吨货物。在技术上，它已经达到了 L4 自动驾驶级别，可以实现特定环境下的自动驾驶。一旦交通环境复杂或天气状况不佳时，它便会切换为人工驾驶模式。当然，为了保证运行的安全，这辆卡车目前的公共驾驶范围为9.5千米，与社会车辆交汇的路段仅有100米左右。

我之前提到，机器学习技术难以被信任的原因是我们无法洞悉它的工作原理，整个系统处于黑箱之中，运行模式也是简单的"端到端"。在现有的认证标准下，我们难以找到强有力的证据来证明它值得信赖。而由于认知论的复杂性问题，在未来的几十年中，这种信赖也难以产生。

如果想要解决这个问题，有两种可能的途径。一种是提高认证标准的门槛。当然，标准的提高可能会阻碍人工智能技术的发展，也会影响自动驾驶汽车的研发。而另一种途径是降低评判标准，比如像美国那样，采用"自我认证"原则。由于没有独立机构的认证，这样的方式完全依赖于制造商的自我管理。至于到底要选择哪种途径，取决于各国政府的判断。

从万物互联，看智能系统如何升级

信风智库： 从智能化的角度看，万物互联将带给我们哪些挑战，我们又该如何应对？

约瑟夫·希法基思： 万物互联是一个雄心勃勃的技术愿景，它的作用在于提高技术的效率和可预测性。这主要体现在两方面：一方面，通过统一的网络部署，让各种基础设施能够被远程感应和控制；另一方面，将现实世界与虚拟计算机系统进行有机整合。

整体上看，万物互联主要由两个部分组成，它们所要实现的目标大相径庭。

第一部分是人类物联网，它为用户提供智能服务的同时也在不断提升互联网技术。更具体地说，其核心愿景是走向所谓的"语义网"。"语义网"是蒂姆·伯纳斯·李在1998年提出的一个概念，试图通过给全球信息网上的文档添加能够被计算机所理解的语义——"元数据"，使整个互联网成为一个通用的信息交换媒介。

不同于传统搜索引擎的"单词式"检索，语义网更强调自然语言的表述。例如，我可以问："在预算1万元的标准下，5月份最佳的海边度假地点是哪里？"

随后，搜索引擎将提供一个选项列表作为答案，这些选项将按照匹配程度进行排序。需要强调的是，尽管专家们在研究和开发方面作出了巨大努力，但"语义网"的这一愿景在很大程度上仍未实现。

万物互联的第二部分称为工业物联网。其核心思路是在复杂工业体系中逐步取代人类，从"自动化系统"转向"自主系统"。这是一个非常具有挑战性的变革。如今，我们已经非常熟悉自动化系统，它能根据设计要求在一定程度上控制设备、电器乃至车辆的运行。

自主系统则非常不同，它们能够在没有人类直接干预的情况下进行长期工作。自主系统的目的是代替人类，需要应对非常复杂的网络物理环境，它们必须同时处理许多不同的目标，而且还要与部分人类合作。当前主要用于智能运输系统、智能电网、智能农场和智能工厂等场景。

目前，工业物联网主要面临两个问题。

第一个问题是网络基础设施与系统的可靠性问题。目前，大部分工业物联网系统可靠性较低，甚至达不到最基础的安全与保障。时至今日，随着连接设备的不断增加，工业物联网系统所面临的风险也越来越高，我们尚未探索出如何保证大型异构系统安全性的办法。

第二个问题是分布式系统的安全性问题。在当前的技术水平下，受制于认知和计算的复杂性，我们很难保证地理分布式系统的安全，除此之外，还需要考虑使用人工智能技术带来的额外问题，因为我们无法保证它的决策时刻符合人类的预期。

总的来说，尽管人工智能、5G和大数据等新一代信息技术在很多领域都取得了重大进展，但迈向万物互联的道路将比我们预期的更长，科学技术的突破能力将决定这项事业的最终结果。

信风智库： 结合您的研究，您认为弱人工智能迈向强人工智能还需要解决哪些问题？

约瑟夫·希法基思： 我想先强调一下弱人工智能和强人工智能之间的区别。

弱人工智能的作用在于解决运算复杂、重复性较高的任务。比如，打造一个围棋训练系统或者高阶方程计算系统。在这样的特定场景中，弱人工智能的表现一定优于人类，因为系统的规则和定义是简单且重复的。可一旦改变应用场景，系统便会失去作用。

强人工智能的目标是接近人类智能，其特点是能够管理许多属性不同的潜在冲突目标，包含短期目标和长期目标。如家庭管理、职业规划和社交沟通等复杂的任务，要求系统必须具备预测外部环境的能力。

虽然只有一字之差，但弱人工智能和强人工智能之间存在着很大的差距。对于弱人工智能来说，计算是唯一的复杂性问题，而对于强人工智能来说，前文提到的所有复杂性问题都会涉及。

20世纪七八十年代，强人工智能的研究者们发现，实现人工智能的认知和推理是难以跨越的障碍。于是很多科学家和工程师们转向了更加实用的、工程化的弱人工智能研究。毫无疑问，他们在这些领域取得了丰硕的成果：人工神经网络、支持向量机……甚至最简单的线性回归理论在足够大的数据量和计算量支撑下，都可以获得令人满意的结果。这些技术变成了如今的人脸识别与语音识别等常见的功能，可即便如此，弱人工智能依然和强人工智能差得很远。

那么，从弱人工智能到强人工智能需要解决哪些问题呢？

我认为主要有两点：

第一个问题是，今天的机器在"情况意识"方面无法超越人类。人类拥有"常识性知识"，这些知识是在生活中逐步建立的。常识性知识代表储存在我们记忆中的概念和事件之间的关系。这些关系通过日常训练逐步建立，无需特殊的学习过程。

举个例子，给一张木头和火柴的照片，人类会自然而然地联想到"火"。但这对于机器而言并不容易，因为它们缺乏这种常识推理能力。迄今为止，常识推理能力已经困扰了人工智能领域近50年。

2019年，知名人工智能研究机构OpenAI发布GPT-2语言交流系统，这个具有15亿参数的通用语言模型一时引发了巨大的轰动，还曾被《经济学人》杂志作为对象"采访"过。该模型生成的句子流畅度惊人，几乎能以假乱真，以至于OpenAI公开表示，因担心它太过优秀带来社会安全隐患而没有将模型完全公开。

随后，一位测试人员在GPT-2中输入了"当你把引火柴和木头堆在壁炉里，然后往里面扔几根火柴时，你一般是要……？"如果这套系统足够聪明，那么它一定会回答"生火"之类的词语，但事实却是一段毫无逻辑的错句。这种对世界的基本知识进行推理的能力，几十年来一直都是人工智能领域难以逾越的大山之一。

倘若要让人工智能具备相同的能力，那么它们必须拥有一个常识性知识的数据库。大部分常识类知识具有隐式属性，使得相关信息难以被明确表示出来。虽然早期研究者认为，可以通过把现实世界的事实都记下来构建一个知识库，并以此作为实现自动化常识推理的第一步。然而这种做法实现起来远比听起来难。

第二个问题是如何使人工智能学会处理"新问题"。瑞士植物学家让·皮亚杰说："智力不是你知道什么，而是当你不知道时会做些什么。"当遇到一个新问题时，人类懂得如何建立一套完整的解决方案，并通过举一反三的能力解决新问题。

最近几年，人工智能也开始探索从单一向综合转变，它需要系统同时具备模块识别和决策智能的能力。模块识别方面，单一的人工智能系统已经能够达到 90% 左右的准确性。然而，在决策智能方面，我们需要将模式识别的结果转化为现实世界中实实在在的决策。通常来说，这种决策并不是相互独立的，而是需要同时对多个问题做出决策，甚至还有可能伴随着较高的风险。正因如此，人工智能的目标也发生了本质变化，不再只是研究如何在单个计算机上重现人类智能，更重要的是如何构建现实世界中的系统，从而解决现实世界中的超大规模问题。简而言之，人工智能正逐步由原理研究走向人工智能综合工程。

这种认知的改变，与传统的人工智能研究的目标有很大差异。当前，人工智能更多的是解决交通、医疗、应急和金融等全球范围内的超大规模问题，而不再拘泥于下象棋等初级型态。

创新生态系统的三大要素

信风智库： 智能时代，技术对创新的驱动作用越来越重要。我们应该如何构建一个与之匹配的创新生态系统？

约瑟夫·希法基思： 创新是现代经济的推动力，是科研成果在新产品和服务开拓中的应用。创新生态系统的概念是在20世纪最后几十年出现的，是为了加速知识的生产和转让，以便用于开发产品和服务。我们今天所说的创新生态系统是指维持创新的重要结构，该系统是三类参与者协同作用的结果，即大型企业、学术机构和初创企业。

第一个是学术机构，它代表各类研究机构和大学，主要任务是培育优秀的科学家和工程师人才，并开展前沿领域的科学创新研究。

第二个是科研项目，它指的是大公司与学术机构展开合作，解决领域和行业内面临的共同挑战。

第三个是初创公司，创业公司体量小、灵活性高，能够在最优条件下实现研究成果的商业化价值。

在创新生态系统中，这三个要素之间具有显著的协同作用，三类参与者都扮演着特定的角色：大企业提供资金并提出新问题；学术机构提供基础研究技能与知识；最后，初创企业能够高效地实施创新过程。

我想强调的是，创新没有独特的秘诀，也没有独特的模式，并非只有经济巨头才能占据令人羡慕的地位或者发挥领导作用。重要的是那些卓越的科研人员、突破性的创意，以及将这些转化为高科技产品和服务的能力。这些都比资金和物质资源更为重要。

发展创新型经济，需要坚持不懈的努力，涉及多种结构性变革，不能简单地将其归于经济问题。仅靠出台条例、法令和法规无法实现创新。创新应为各个利益相关方协同合作下的产物，这些利益相关方将成为执行提案的主要参与者，例如企业、研究机构与金融机构。

在此，我想强调创新过程中人的因素，因为在现有政策和实践中，人的因素常常被忽视。我认为尤其重要的是建立将学术研究与实体经济联系起来的措施和激励机制。学术机构需要达到临界规模，一个创新项目必须积累大量多学科专业知识才能取得研究成果。此外，应充分认可研究人员的创造力与活力，并创造有利条件，吸引全世界最优秀的研究人才。

如今，尤其是在大城市，人类面临诸多全球性挑战，如环境挑战、能源挑战、人口增长挑战、安全挑战、医疗和教育挑战。创新以及创新成果的合理应用，对应对这些挑战、实现社会福祉和繁荣至关重要。

贺东东：
工业互联网如何重塑未来制造

贺东东　树根互联股份有限公司（以下简称"树根互联"）联合创始人、CEO，工业互联网产业联盟副理事长，华中科技大学兼职教授。曾任三一集团高级副总裁兼首席流程信息官，三一印度有限公司、三一德国有限公司董事长，并登上2020年度AIoT产业领袖人物榜，荣获2016年度中国信息化建设杰出CIO、2015年《IT经理世界》全国杰出CIO TOP5。

树根互联是专注打造工业操作系统的平台公司，旗下的根云平台可以面向机器制造商、设备使用者、政府监管部门等社会组织，在智能制造（透明工厂）、智能服务、智能研发、产业链平台、工业AI、设备融资等方面提供数字化转型新基座；不仅是国家级"跨行业跨领域工业互联网平台"，同时也是首家入选Gartner IIoT魔力象限的中国工业互联网平台企业。

截至目前，树根互联已为超过60个国家和地区、81个工业细分行业提供服务；还通过"通用平台+产业生态"的P2P2B模式，结合行业龙头企业、产业链创新企业等生态伙伴的行业经验和应用场景，形成20余个产业链/产业集群平台，覆盖工程机械、混凝土、环保、铸造、塑料模具、纺织、定制家居等多个产业。

扫码观看
访谈精选视频

我们如何认识工业互联网与智能制造间的关系？工业互联网将带给制造业哪些深刻的改变？一个成熟的工业互联网平台，需要具备哪些能力？工业互联网如何为产业集群的发展赋能？

为什么说工业互联网是新型基础设施

信风智库：工业互联网与智能制造之间，是一种怎样的关系？我们如何来理解认识？

贺东东：智能制造其实是对制造业终极状态的描述。当数字化转型实现以后，制造业就会达到一个具备智能化的制造水平，我们称这个阶段为智能制造，它是一个更加宽泛的概念。如果从实现路径上来讲，智能制造其实涵盖了所有工业生产要素的进步与升级，包括硬件、软件、工艺、材料、能耗与机器等。

工业互联网是实现智能制造的一个关键核心技术，也是智能制造最重要的基础设施。也就是说，制造业如果想要达到智能制造这样一个终极状态，就必须依靠工业互联网技术去推进，并且依靠工业互联网平台去承载与赋能，这就是二者之间的关系。

为什么说工业互联网是智能制造的关键技术呢？不论是智能制造还是工业4.0，其核心都是数字技术。数字技术一方面要将工业的物理对象数字化，另一方面又要基于数字化对象融入新一代信息技术。工业互联网一方面把工业体系中人机料法环各种生产要素变为各种数字化的对象，另一方面结合大数据、人工智能以及云计算等新技术，驱动已经被数字化的各种生产要素自动化工作与智能化运转，从而达到智能制造的状态。这就是工业互联网实现智能制造的技术逻辑和载体价值。

此外，我们从存量技术和增量技术的角度来看，工业互联网又是智能制造的必要补充。全球的制造业发展从工业1.0到工业3.0，其实已经创造与沉淀了很多信息化与自动化的制造技术，比如ERP与MES等工业软件，还有自动化生产线和工业机器人等硬件，这些都是存量技术。而如今工业4.0时代的核心是以数字化驱动智能化，其具体形式就是以大数据、人工智能与工业互联网为代表的增量技术。但这些增量技术并不能颠覆存量技术而独立存在，它们必须叠加在存量技术之上，二者结合从而实现智能制造。其中，工业互联网既是技术能力，又是载体平台。从这个角度看，工业互联网也可以说是实现智能制造的最后一块拼图。

信风智库：从未来着眼，工业互联网将带给制造业一种怎样的深刻改变？体现在哪些具体方面？

贺东东：工业互联网对于制造业而言，主要解决了三个层面的问题，即制造业的数字化、网络化和智能化。

首先是数字化。为什么我们制造业的数字化相对互联网上半场的消费互联网而言还较为落后？主要原因在于制造业跟钢铁打交道，还停留在针对物理对象的作业上。如果物理对象没有变为数字对象，就不能够跟日新月异的新一代信息技术连接上。而工业互联网平台则填补了这个空白，它将物理对象数字化，建立一个"数字双胞胎"或"数字孪生"的模型，真正意义上将制造业的物理资产变成数字资产。

其次是网络化。当制造业所有相关的物理生产要素变成了数字要素后，工业互联网就可以把它们全部连接起来，使得机器与机器之间实现互联，这就能形成更大范围的机器协同与共享，让所有工业生产要素在同一个平台上统筹运转。

最后是智能化。在数字化和网络化的基础之上，制造业与新一代信息技术融合就成为可能。我们通过大数据、人工智能、云计算与区块链等新技术去优化工业互联网构建的数字模型，优化后的数字模型反过来又能更好地指导我们提升制造业生产效率，并开发更多的商业模式。这时，智能制造就在一定程度上实现了。

实际上，工业互联网真正意义上的价值是打通了制造业物理世界、数字世界和新一代信息技术之间的鸿沟，它起到了新型基础设施的作用。为什么说工业互联网是基础设施呢？我们从两个方面来看。

一方面，基础设施有一个很重要的特点，就是它具备公共属性且工程量巨大。它不是某一个用户自己就能够建造的，它是一个公共的巨大投入，然后大家一起来共享使用。

工业互联网也具备这个特点。中国制造业发展参差不齐，很多中小型制造企业信息化水平不足，自身又不具备强大的财力、物力与IT技术，无法打造工业互联网体系。所以这就需要更强大的平台或机构，利用更强大的技术与资源，打造工业互联网平台提供给中小型制造企业使用。

另一方面，当基础设施建立起来后，大家在享受公共设施带来便利的同时，花费的成本是非常低的。比如高铁这样的基础建设需要极高的投入，但我们乘坐高铁的费用却在可承受的合理范围之内。

同样，工业互联网也是具有普惠属性的。因为工业互联网是集中与沉淀了大量资本、技术以及资源的平台，它几乎涵盖了制造领域内企业的所有需求，而企业可以按需付费，以制定服务的方式去参与和使用，门槛低且成本少。例如，以前一个传统的制造型企业要开发一款 ERP 软件，可能要几百万元，而在工业互联网平台上，购买一个同样功能应用的接口，可能只需花费几万元，这样就用最低的成本享受到了最新的技术与服务。

工业互联网就像一条制造业领域内的信息高速公路，为制造企业带来了新技术的巨大便利，且成本很低。所以我们可以认为工业互联网是新型基础设施。

信风智库：一个真正意义上成熟的工业互联网平台，需要具备哪些能力？

贺东东：一个成熟的工业互联网平台至少要具备五个能力。

第一个是广泛且可靠的设备连接能力。一方面是广泛，我们对智能制造的要求是万物互联，将大大小小的制造型企业与各种各样的机器设备都连接到平台上来，这里面涵盖的机器设备种类就会特别多，相应的协议控制指令集也会特别多；另一方面是可靠，在大规模的工业生产体系中，机器设备每时每刻都在高速运转，必须保持工业级别的稳定，不能出错。

第二个是工业数据的处理能力。这也可以从两个方面来谈。一方面，要算力强大。工业数据的特点是海量、高频和高并发，数据处理的时间单位是毫秒。几毫秒就会有海量数据同时出来，而且在这么短的时间内还要处理、判断以及决策，这就对数据链路的要求特别高，还得采用流式计算。这一点跟自动驾驶很相似，机器必须在毫秒级的时间单位上判断出了什么状况，并马上停机，甚至自动进行预防干预。

另一方面，要具备深刻的工业认知度。不论是大数据、人工智能还是云计算，如果技术不能深入行业，不能理解制造业的运转机理，就无法构建出符合工业体系的数字模型，也就解决不了制造业的实际问题。举个例子，我们建立的数据模型必须跟工业机理吻合。一个机器可能分了好多层，有某种特定的架构，不同模块之间存在不同的内在逻辑，这些都要反映在数据模型里面。

第三个是开放的应用开发能力。制造型企业的核心需求非常多，有些需要平台帮忙做后市场管理，有些需要实现智能化生产提升效率，有些需要平台提供商务和金融方面的服务，所以工业互联网平台既要具备很强的包容能力，让各类工业应用软件能够互相打通数据和兼容集成；又要具备足够的开放性，让不同的开发者针对不同的客户需求基于平台开发不同的应用软件。

其实工业互联网也是工业领域的操作系统。就像移动互联网各个开发者都可以基于安卓这个操作系统，开发游戏、购物和音乐等不同类型的 APP 应用软件。

第四个是通用型应用的沉淀能力。工业互联网服务不同类型和不同领域的制造企业，可以沉淀出一些具有共性需求的通用型应用，形成一个即插即用的服务模块，其他制造企业不用从零开始，这就大大提升了应用开发效率。就像手机应用商店沉淀了大量 APP，工业互联网平台沉淀的工业应用越多，则意味着企业实现数字化转型的成本越低、速度越快，平台价值也就越大。

第五个是生态伙伴的连接能力。制造业是一个庞大的生态系统，没有哪一家企业能够覆盖所有的需求。所以需要借助工业互联网的平台来整合制造业产业链上下游，实现生态伙伴的连接。这才是真正成熟的工业互联网平台。

数字化与智能化之路该怎么走

信风智库： 数据是智能制造的基础。在采集、整理、分析、应用与治理等环节上，工业大数据与消费互联网有哪些不同？

贺东东： 工业互联网被称为"第三张网"。以前有很多人将工业互联网理解为工业加消费互联网，所以我们看到有一些互联网公司跟制造型企业签协议，说一起来做工业互联网，但最后都没有办法解决问题。原因在哪里呢？

其实在大数据的层面上，工业互联网与消费互联网有很大的差别。第一个是数据采集方式不同。消费互联网一般是通过标准化的机器采集数据，比如手机与计算机等设备，虽然面对的是千人千面的用户，但人的共性是一样的，比如采集用户的消费行为，看了什么网站，下了什么订单等，相对而言是标准化地采集。而工业互联网连接的则是成千上万种不同类型的机器设备，这些机器设备内部的各个模块之间、设备与设备之间的机理是完全不同的，比如机床与机器人就截然不同，背后的用户行为与操控逻辑都不一样，所以工业数据的采集一般是多源异构且标准不统一的。

第二个就是数据量的差异。因为工业机器是时时刻刻持续运转的，所以它是高频且全面地长时间采集数据，包括温度、压力、规格与能耗等工艺参数，这就意味着数据量巨大。就像自动驾驶汽车持续行驶,当中产生的加速、减速、转向与油耗等行车指令数据量是非常庞大的。

第三个就是应用重点不同。相比于消费互联网,工业互联网更注重实时采集,快速处理决策。因为工业互联网的核心是机器互联,工业数据的价值在于能够指导生产、预判风险和提升效率。试想,如果不能及时通过工业互联网进行数据处理分析,对故障风险做出预判,后果就会很严重,哪怕一秒钟、一毫米的微小误差,也会出现故障导致机器停机,影响整个生产线的生产效率。

这当中其实还涉及流式计算的问题。工业体系中的机器数据采集是高频且动态的,所以要求工业互联网能够在大规模、不断变化的流动数据运动过程中实时地进行分析,捕捉到可能有用的信息,并把结果发送到下一个计算节点,从而指导机器高效运行。比如柴油机跟液压系统之间的配合,如何通过前一个环节的指令来调整后一个环节的指令,数据模型必须理解并且控制这些机器动作,这是非常重要的能力。

这就要求数据模型深刻地理解工业机理——某个数据代表了哪种控制指令,这个控制指令是什么意思,会导致一个什么样的机器动作;各个部件或设备之间在工艺流程上是制约关系还是呼应关系。所以说,高频采集、海量数据、具备流式计算能力与懂得工业机理的工业互联网,跟消费互联网是完全不一样的逻辑。

虽然,工业互联网和消费互联网都叫互联网,二者在基础技术方面都是共通的,比如存储与计算等,但是工业互联网和消费互联网在构建数据模型时的逻辑完全不同,使用的数据链路也完全不同。这也解释了为什么用消费互联网的逻辑来打造工业互联网,以此解决工业问题是行不通的。

信风智库：新一代信息技术与工业的融合过程中，普遍需要解决哪些核心问题？

贺东东：举一个中医药发展的例子。中医药起源于中国，经历了千百年的积累和"师父带徒弟"式的传承，比如各种草药的药效研究以及大量的治病案例。自古以来，越老的中医水平越高，因为老中医长时间沉淀了丰富的药理知识与病例经验。但是如今却发生了一些变化。我们看到日韩等国家，利用数理与数据对中草药的成分进行分析，用新的技术去掌握药材的作用机理，形成了新一代的中成药体系，他们的中成药发展水平已经赶上了中国。

这反映出了什么问题？一个是新一代的数字化路径，一个是千百年来"师父带徒弟"的传统方式，显然数字化的发展路径比传统方式速度更快、效率更高，而且还更容易形成体系化的增长。实际上，诺贝尔生理学或医学奖获得者屠呦呦也是在大量搜集民间中医药方的基础上，运用数理分析的方式发现青蒿素可以治疗疟疾的。

回到工业上来，其实欧美发达国家的工业优势，跟中国的中医药发展很类似。他们在工业 3.0 阶段积累了大量的工艺技术，很多企业都只专注于某一个技术，比如数控机床的加工制造，然后基于这一技术不断优化材料、工艺和人才，不断完善与沉淀技术，类似于老中医"师父带徒弟"的方式进行积累与传承。所以我们看到很多欧美制造企业在某一细分领域都做到了全世界领先。

而工业 4.0 则带来了数字化发展路径，提供了新的技术积累方式。那么新在哪里呢？比如当我们需要用数控机床去加工一个零部件的时候，工艺其实可以用数据模型来表达，具体而言就是调整设备的参数，在哪里切，用什么刀切，切多薄多厚，如何调整转速等，这些都是欧美国家长期以来的技术积累。以前我们即使拿到欧美制造企业的工艺图纸反向测量，或者使用一样的甚至更先进的机床，但加工出来的零部件在精度与质量上，就是不如别人。这就像老中医抓药方子，每一方药配什么种类的药材，配多少量，都靠老中医的经验。

但是在如今的智能时代，我们就能构建数据模型将长时间积累的经验用工艺标准表达出来，然后通过多次生产，将质量结果与欧美标准进行反复对比，同时采用人工智能反复学习训练，不断优化工艺参数，最终就能达到甚至赶超欧美的先进经验水平。这就像AlphaGo下围棋，人工智能快速学习海量棋谱后，就能轻松打败人类高手。

其实，这也就是工业3.0的存量技术与工业4.0的增量技术相比的一个差异。虽然中国制造业整体而言还达不到工业3.0的阶段，但是我们如果掌握并运用新一代信息技术带来的全新学习方法与进步路径，就能缩短差距甚至实现赶超。

那么这一过程中的困难在哪里呢？其实核心还是在数字化技术与工业机理的结合上。懂工业的人不懂技术，懂技术的人又对工业认知不深，这也是像我们树根互联这样的工业互联网公司面临的主要挑战。

最直接的体现就是在人才上。我们要让互联网人才去理解制造业的内在逻辑与机理，然后将这些行业认知运用到平台软件开发上；反过来，如果是制造业相关的人才，我们就要让他们去了解新一代信息技术，比如数据建模和人工智能，以及平台工具该如何构建。总的来说，就是要把互联网人才和制造业人才揉在一起，让他们在工作过程中彼此交流学习，融合形成复合型人才。

而进一步体现在产品上,就是要构建真正符合制造业的数据模型逻辑。我们将这种核心技术称为"物模型",就是通过一个相对统一和通用的数据模型,既能够针对各种各样的机器进行建模,又能够兼顾制造业的内在机理,还能让软件工程师基于这个模型去做各种各样的应用。所以,在人才与产品上,将互联网与工业融合在一起是最核心的难点,也是最有价值的地方。

信风智库: 从市场角度,结合实践案例,在您看来,如何让工业企业尤其是中小规模的工业企业接受工业互联网等新技术?

贺东东: 工业企业针对数字化转型一般分为三种态度。第一种是不想转,这类企业一般对目前现有的行业、市场以及模式相当熟悉,它们不愿意去冒风险转型升级;第二类企业则是不敢转,这类企业比较谨慎,对很多新兴技术持有怀疑态度,不确定是否能为自身带来价值;第三类企业则是不会转,它们虽然想转型,但是不知道怎么去转型,也没有数字化人才的支持。

其实从供需两方面来看,归根结底就是缺钱。对于制造企业来说,自动化、数字化与工业软件的好处以及数字化转型的大道理,大家都懂。但是制造业大多数赚的都是血汗钱,大家没有那么多钱投入数字化转型。然而,制造企业数字化转型这一步,越是迈不出去,就会越来越被边缘化,进而越来越没钱,形成了恶性循环。

反过来,对于我们这样的工业互联网供应方来说,核心就是要打造性价比很高、看得见摸得着的应用。因为即使再怎么没钱,如果投入 10 块钱就能赚回来 15 块钱,这种看得见摸得着的投入产出,制造企业肯定会干。对于一家制造企业来说,问题很现实：能不能把成本降下去，能不能多卖几台设备，能不能把能耗降低 10%。如果能实现，那么要花多少钱？只要算得过来账，这个事情就肯定可以做。

作为新型基础设施的工业互联网，就成了最有效率的解决方案。我们投入了上亿的资金和大量的人才，来打造这样一个技术平台，在以 SaaS 模式推广的时候，制造企业只需要花很少的钱就可以用上新技术。

比如我们有一个做果蔬烘干机的中小企业客户，如果一开始就要他们花几百万元来做一个智能化平台，他们肯定会犹豫或者直接拒绝。所以最初只让他们花两万元，连上 10 台烘干机，这种程度的尝试，他们完全可以接受。在疫情期间，因为使用了工业互联网平台，在没办法派人出差的情况下，通过远程在线诊断故障，远程指导当地电工维修，他们维护了好几台烘干机。以前，还要派人出差，响应速度也较慢，现在不仅节约了差旅成本，还通过远程指导本地服务，加快了响应速度。这就是看得见的好处。如今，这家果蔬烘干机企业不仅全面地用上了工业互联网，还开发出了全新的商业模式。以前他们只是把烘干机卖给果农，现在他们研发了车载移动果蔬烘干机，直接把车开到各地果蔬农场，把收购来的新鲜果蔬就地烘干加工为果干，然后再以高附加值向市场上销售。一个靠卖设备的制造型企业，依靠工业互联网形成了一种打通一、二、三产业的新商业模式。

我们可以看到，工业互联网就是一种集中投资建设，低价普惠使用，还能产生各种个性化应用价值的新型基础设施。我们经常比喻：旧时王谢堂前燕，飞入平常百姓家。以前只有大型企业才能每年花几千万元，组建上百人的 IT 工程师团队，自己做信息化与数字化改造，如今中小企业花很少的钱，也能拥抱工业互联网带来的智能时代。中小企业占据了中国制造业至少半壁以上的江山，让新一代信息技术能够普惠与普及广大中小型制造企业，工业互联网这种载体与模式是非常合适的。

信风智库：对于工业互联网的发展，标准体系的建立意味着什么？在实施过程中，这项工作面临哪些难题需要解决？

贺东东：从工业互联网发展的角度出发，建立标准是最重要的一件事。我们知道工业互联网的一个关键能力，就是广泛的机器连接能力，但是其中的挑战在于，机器种类太多，每一种机器都有自己不同的协议与指令。

我们买手机，不管华为、小米还是苹果，一买来就可以上网使用。为什么？因为手机的入网标准、通信标准是一样的，三大运营商的指令集是一样的。无论是哪家厂商的手机，插上SIM卡就都可以使用。

我一直在呼吁一件事情：我们制造业如果能够把所有机器设备都当成手机一样，当成一个网络设备，然后统一通信标准和网络的指令集标准，这样就能让我们整个工业数字化服务的成本大幅降低，降低到原来的1%都有可能，工业互联网的普及速度也会大幅提升。作为工业互联网平台，我们也就没有必要制造那么多通信盒子，然后花费那么多人力与精力一台设备一台设备地去接入网络了。

其实，制造业的机器已经是一个网络终端了。如果反过来看，工业互联全部实现，所有的机器都联网，不就是一个个网络终端了吗？对于机器设备的生产厂商而言，需要统一的是通信标准和外部指令集，而内部元器件和构造都可以保持不一样，这跟各个手机厂商也有不同的内部元器件与构造是一样的。这样就不会剥夺设备厂商的差异化竞争力。

我们再从工业互联网平台的层面来看，任何一家平台都不可能把全天下的设备都连上来，所以不同平台上的制造企业与设备需要具备跨平台协同的能力，这就需要我们建立跨平台之间的数据交换标准和数据接口标准。这样才能更大范围地去普及工业互联网和智能制造。所以，为广泛连接机器和更大范围应用这两个角度，建立标准是非常有必要的。

工业互联网将给产业带来哪些改变

信风智库：在区域经济发展的层面，很多地区都重视产业链招商，那么工业互联网如何为产业集群的发展赋能呢？

贺东东：产业集群是中国经济发展的一种重要模式，我们可以看到很多区域，从县到乡甚至到村一级都聚集着同一种业态，比如有些地区专门做小家电，有些地区全部做纺织品等。这些产业集群其实都具备两个明显的特点：

一是它的总体规模巨大，但单个企业规模较小。我们经常看到一个区域所有的企业都在做同一件事，产业集群能够做到几十亿元甚至几百亿元的规模。但是具体看某一家企业，它的个体规模又都不大。这就像一个蚂蚁兵团，它们集合在一起能够形成非常大的力量，但单只蚂蚁的实力却有限。

二是产业集群的业态一般都是集中在产业链条的某一段：要么都是生产某一种东西；要么都是销售某一种东西，做这个东西的贸易环节。

这两个特点带来了什么问题呢？现代产业竞争都是产业链的竞争，从订单到生产，再到物流，最后到服务，每一个环节都需要企业具备优势，它是综合竞争力的体现。如果只是在产业链的某一段具有优势，在资讯发达、市场变化日新月异的今天，就会造成市场反应落后的问题。举个很简单的例子，即使产品工艺和质量再好，但是没有及时打开渠道，销售就无法跟上，市场价值也会大打折扣。反过来即使销售做得再好，代理的产品工艺和质量没跟上，也不能长久。

那么，工业互联网如何针对产业集群进行赋能呢？

第一种方式是整合需求共享赋能。我们把某一个产业集群的共性需求整合在一起，通过工业互联网平台赋能，补齐大家共同需要的能力。比如广州定制家具产业。我们知道定制家具已经是一个大趋势，用户根据居住房屋的面积与布局，定制尺寸与之匹配的、形状颜色符合个性化需求的家具。那么，定制家居需要具备什么能力呢？它需要通过数字化的方式，将前端的设计软件、中端的生产制造软件以及后端的物流与安装服务全部打通，让前中后各个环节高效协同地推进流程，才可以实现。然而，目前在广州的定制家具产业集群里，只有少数大型企业具备这样的能力，规模只占整个产业集群的15%。剩下85%的中小企业不具备这种能力，也没有实力来支撑一个平台。这就需要工业互联网通过平台化的方式，将设计、制造、物流和安装等各个环节进行数字化打通，再以SaaS化的方式提供给各个中小企业使用。中小企业们只需付一点服务费就能拥有跟大企业一样的数字化定制能力了。

第二种方式是匹配需求快速反应。比如我们在湛江打造的小家电平台。湛江生产制造电饭煲和热水壶的能力非常强，但如今互联网消费趋势变化非常快，等厂家做好设计，开发完模具，再制造出来，可能消费偏好就已经变了。所以，湛江的小家电制造厂商需要更敏锐地把握消费需求，需要更新颖时尚的产品设计能力，以及快速开模、快速制造的能力。我们打造的工业互联网平台，就整合了非常优质的设计师资源，以及快速开模的模具厂，通过信息化与数字化打通了从需求到设计，再到开模与制造以及销售的全流程。不少湛江的小家电厂商，都能够通过平台低成本地实现对消费趋势的快速反应。

信风智库：从 M2C 的标准化大规模制造，转变到 C2M 的个性化、离散型、规模化制造，工业互联网在哪些方面提供支撑？

贺东东：要实现 C2M，就要将信息流、订单流、生产流、物流与服务流"五流合一"，让整个产业链形成一个流动的整体，其中就包括门店、设计、制造、销售和服务等整个产业链上下游各个环节的协同。这就需要一个平台和机制来支撑，才能把产业链上各个环节与各个企业串联在一起。

传统的工业软件只能解决某一个环节的问题。比如 ERP 系统，就是企业资源计划，它默认是以企业为服务主体，在企业内部完成资源调度，是有内外边界的。而工业互联网天生就是一种互联网化的平台机制，一开始就把整个产业链当作服务对象，能够打破企业内外边界，让产业链上所有企业能在同一个平台上去协同生产，比如订单的协同、设计的协同以及生产的协同。

当然此前也有一些订单打通与匹配的模式，比如 C2B，但这也只是将消费者与贸易企业或者生产企业里面的销售部门打通，而并没有直接跟制造环节打通，所以只能实现将订单集中起来进行传统的规模化与标准化的批量生产。

真正意义上的 C2M 个性化定制能力，一定是要在制造环节体现出来的。所以"五流合一"就必须通过广泛连接机器的工业互联网，彻底打通制造环节，从而实现 C2M，即消费者到制造端的个性化智能制造。

信风智库： 从"卖产品"到"卖服务"，可以说是近年来制造业在市场端的升级趋势。工业互联网在哪些方面促进了这样的转变？

贺东东： 从"卖产品"到"卖服务"，这一定是个大趋势。从本质上来讲，一个企业购买一台机器，目的其实不是买机器，而是买服务。举个例子，一个企业买了一台挖掘机，它并不是需要挖掘机，而是需要用挖掘机来挖土，最终目的是挖土。就像我们买车其实是为了出行，未来有了更多的智能出行服务随喊随到，我们就没有必要花几十万买车了。

我们把产品价值一直往后延伸，就会发现最后都是服务价值。所以，制造业一定也是服务业。因此我们提出了"即服务"的模式，就像我们平时常说的 PaaS 和 SaaS，平台即服务和软件即服务，制造业通过工业互联网也可以实现"产品即服务"与"运营即服务"，让客户直接享受到机器产品带来的服务，而不再需要购买机器产品这个中间过程。

那么在这个过程中，工业互联网起到什么作用呢？机器如果要提供服务，就必须实现远程可控和可计量计价，不然服务就没有办法形成交易。比如网约车打车平台，通过网络把车连起来，就能计量这辆车从起点到终点跑了多少路程，应该向乘客收取多少费用，而且还能监控整个过程的服务质量。同样的道理在制造业，工业互联网把机器设备连接起来，就能知道某台设备生产了多少零部件，生产质量怎样，谁在使用这台设备，进而就能够对这项生产服务进行计量计价交易。

还有一个重点是之前讲到的 C2M 模式。工业互联网让"五流合一"打通了制造环节，实现柔性化制造，为 C2M 模式提供了强有力的支撑。在这种情况下，一种个性化服务的需求，就能直接转换为制造环节可以生产与运营的一个订单。反过来看，制造环节就可以省去中间环节，更直接地为最终的用户提供个性化服务。

所以，在整合全产业链的基础之上，工业互联网的核心能力在于将机器变成服务，以及将制造变成服务，从而真正推动制造业向服务业发展。

探索智能制造的中国逻辑

信风智库： 在全球争相发展智能制造的大背景下，在发展路径、模式与目标上，中国的智能制造较之欧美国家的工业 4.0 有着怎样的不同？

贺东东： 先说相同的地方，无论是工业 4.0 还是工业互联网，本质上都是对制造业进行数字化，形成数据资产，然后把新一代信息技术运用到数字化的主体上，通过"数字双胞胎"或"数字孪生"的方式，全面优化制造业的运营过程和商业模式。这个是它们相同的本质逻辑。

不一样的地方，在于中外工业发展基础和现状还有一定差别。站在全球视角来看，我们一直说中国的制造业发展大而不强。"大"是指中国是目前工业门类最丰富、工业机器数量最多，以及工业数据资源最多的一个国家，这是我们制造业数字化转型的特点和发展智能制造的优势。那么为什么又说"不强"呢？就因为中国目前制造业的数字化水平发展参差不齐。虽然我们也有像三一重工和华为那样数字化水平处于世界一流的龙头企业，但也有很多中小型制造企业几乎没有实现信息化和自动化。所以从整体来看，目前中国的制造业可能还没有达到工业 3.0 的阶段，相对而言还有待提高。

基于这样的现状，中国智能制造的发展与欧美国家的战略路径肯定会有所区别。首先是挑战和需求不一样，我们目前在大多数情况下主要利用工业互联网和数字技术去解决一些基础问题，比如怎样通过工业互联网对基础设备进行运行维护，以及如何精准地匹配劳动力和统计薪酬等；而欧美国家已经在用工业互联网对航空发动机进行预测性维护。虽然技术的本质是一样的，但技术应用的场景和方向不同，创造的价值也就有差别。

其次是想象空间不一样。在大数据时代，数据堪比石油，是最重要的生产资源之一，而中国拥有大量的工业机器和工业制造数据，以及丰富的应用场景。从这个层面来看，中国发展工业互联网和智能制造的潜力巨大。实际上，智能制造的逻辑就是将新一代信息技术通过数字化导入制造业的应用场景中。在这一过程中，中国无疑具有规模上的领先地位。

最后是重视程度不同。从现状来看，中国对工业互联网和智能制造的热情是超过欧美国家的。对于很多制造业发达的欧美国家而言，工业 4.0 只是被视为一个新的技术方向，大家按部就班地推动发展就够了。而中国的智能制造已经上升为国家发展战略，从各级政府，到学术界，再到企业界，都投入了极高的热情，大家都高度重视且积极地推动落地。所以，中国智能制造的未来是非常乐观的。

2020 年工业和信息化部发布的跨行业跨领域工业互联网平台清单

单位名称	平台名称
青岛海尔股份有限公司	COSMOPlat 工业互联网平台
航天云网科技发展有限责任公司	Indics 工业互联网平台
北京东方国信科技股份有限公司	Cloudip 工业互联网平台
江苏徐工信息技术股份有限公司	汉云工业互联网平台
树根互联技术有限公司	根云工业互联网平台
用友网络科技股份有限公司	精智工业互联网平台
阿里云计算有限公司	Sup ET 工业互联网平台
浪潮云信息技术有限公司	浪潮云洲工业互联网平台
华为技术有限公司	FusionPlant 工业互联网平台
富士康工业互联网股份有限公司	Fii Cloud 工业互联网平台
深圳市腾讯计算机系统有限公司	WeMake 工业互联网平台
重庆忽米网络科技有限公司	H-IIP 工业互联网平台
上海宝信软件股份有限公司	xIn3Plat 工业互联网平台
浙江蓝卓工业互联网信息技术有限公司	supOS 工业操作系统
紫光云引擎科技（苏州）有限公司	UNIPower 工业互联网平台

注：以上排名不分先后

信风智库：有一种说法是，智能制造就是机器换人。站在工业制造不断进化的历程上，从实际工艺流程升级的角度出发，我们应该如何理解智能化与从业工人的关系？

贺东东：我们放大时间维度来看这个问题。1000年前没有机器，几乎所有的劳动都是靠人力；再看1000年后的今天，大量的机器在很多领域都已经替代了人力。比如农业，以前全靠人力耕种，而现在可以使用各种机器耕地、播种和收割，大大减少了人力劳动。但事实上，我们看到全世界的就业人口是在大规模增长的，从来没有因为机器的出现而受到负面影响。

这是为什么呢？从两个方面来看。第一是新的需求将创造新的职业。人类需求的增长永远会超过生产力的发展，这也是人类文明进步的源动力。对于社会提供的商品与服务，人们永远会产生各种各样的新需求，那么这就催生了各种各样的新生产力与新职业。

比如工业互联网要把所有机器设备连接起来，那么如何稳定地连接，连接后又怎样将机器数据打通，又如何监管和使用，这就需要大量既懂工业又懂IT的新型人才。所以我们专门开发培训课程，跟职业院校合作，培养这样的人才，这个过程就创造出了大量新的工种。

第二是人类的劳动价值在不断转移。比如在农耕时代，几乎所有的劳动力都集中在农业，当工业时代到来，农耕机器出现，劳动力就从农业转到了工业。虽然农业的劳动力转移了，但因为机器提高了生产效率，农业的产值却不降反升。现在来到了智能时代，工业实现数字化转型与智能制造的过程，机器又将取代传统的生产工人，这时人类的劳动力又会向数字化与智能化领域转移，比如工业数据分析师、人工智能工程师和信息安全工程师等。

近段时间在全球工业与能源领域，我们看到网络安全事件频发，造成了不可估量的损失。这就意味着智能时代催生了大量的信息安全需求。以前一座传统工厂可能需要 20 个保安人员，现在一座智能工厂可能需要 20 个网络保安人员，也就是信息安全工程师。

所以，从人类历史发展的维度看，科学技术从来没有减少就业人口，只会让就业岗位越来越多。

周 涛：
如何打造新一代数据治理体系

周涛 电子科技大学大数据研究中心主任、教授、博士生导师，成都大数据产业技术研究院院长，重庆大数据人工智能创新中心主任、首席科学家，成都新经济发展研究院执行院长。

2005年于中国科学技术大学获学士学位，2010年于瑞士弗里堡大学物理系获哲学博士学位，撰写出版专著6册（《金融复杂性：实证与建模》《链路预测》《个性化：商业的未来》《最简数据挖掘》《重塑》《为数据而生：大数据创新实践》），在 Physics Reports、PNAS、Nature Communications、PRL 国际SCI期刊发表300余篇学术论文，论文被引用超过30000次，H指数达到79。

2011年获第十二届中国青年科技奖；2012年获四川省五四青年奖章；2013年获四川省科技进步一等奖；2014年起历年入选Elesvier最具国际影响力中国科学家名单；2015年当选年度中国十大科技创新人物；2016年获首届"四川杰出人才"奖；2017年获全国创新争先奖；2019年获四川省师德楷模称号。

扫码观看
访谈精选视频

大数据领域的发展正处于何种水平与阶段？大数据如何支撑人工智能与区块链这些新技术的发展？大数据行业进一步发展，还需要解决哪些问题？如何设置数据的标准，如何把控数据的质量？如何实现数据价值化和数据资产化？数据要素市场化流通需要哪些基础条件？我们应该如何应对数据治理与数据安全问题？

如何激活大量沉睡的数据资源

信风智库：大数据的概念从诞生至今已有10多年，其逐渐被大家接受，各种应用也纷纷落地。在一个更长远的时间维度上，您认为大数据领域的发展现在处于怎样的水平与阶段？

周涛：自IBM在2009年提出大数据概念，已有将近13年的时间。总体来说大数据还是处在一个大行业的早期，或者一个方兴未艾的阶段。虽然数据在某些垂直行业已经开始发挥作用，并产生价值，但是，我可以保证地说90%以上的数据，依然没有发挥其应有的价值。

大量的数据资源还处在沉睡状态中。我们应该思考如何甄别哪些数据是有价值的，应该在哪些场景发挥价值，并以一种更自动化、更低成本的方法对数据质量进行控制，从而真正完成数据的价值变现。这个层面的工作，我觉得还没有大范围地开展。

所以，大数据现在大体上还处于刚刚被人们认知的阶段，人们知道它是有价值的东西。从目前的情况来看，很多人仍然不愿意把自己的数据就这么轻易地给别人。但实际上你就算把数据给了别人，别人很可能也挖不出什么价值，你留在自己手上，可能也没什么价值。大数据还处在这么一个早期阶段。

信风智库：在您看来，不管是政府管理者、从业者，还是社会公众，大家对大数据的认识还存在哪些误区？

周涛：目前为止，大家对大数据的认知还是有一些普遍的误区。第一个比较典型的误区，就是总觉得数据要大到一定规模才能产生价值。实际上大数据的大，主要是指价值大。它并不是说我们必须要有一个 PB 的数据才能叫大数据，如果我们能用很少的数据在关键的决策中起到支撑作用，我们也认为它是大数据。

第二个比较大的误区，是认为一切数据都一定会有价值，因此比较盲目地投入大量成本去把这些数据保存下来。在某种意义上，它可能也是对的，因为我们虽然在当下没有认识到这些数据的价值，但也许 10 年、20 年以后能认识到。然而在很长时间内，相当大的一部分数据没有什么显著的价值。那么，去采集存储这些数据，实际上是一种无用的投入。

所以必须分辨清楚，哪些数据是高价值、可流通的数据，哪些价值可能会低一点，哪些我们在很长时间内看不到价值。我们需要认识到，并不是所有的数据都是资源或者资产。

第三个比较大的误区，就是很多人认为，拥有了海量数据之后，可以通过数据的关联分析作出预测与判断，而不依赖于行业认知和因果关系。这实际上是不对的。因为现在没有，未来也不会出现通用数据模型，更不可能依靠一个模型解决行业的一切问题。所以行业认知很重要，并且在大部分需要干预和调控的地方，我们必须要知道因果。只知道关联，可以进行一些预测，但是不能有效干预。

第四个比较典型的误区，就是把大数据和人工智能的核心功能弄混了。认为所有的大数据公司也是人工智能公司，所有人工智能公司也是大数据公司。其实不然，相当一部分人工智能是依赖数据的，但有很多人工智能是不依赖数据的，它可以通过增强学习、自我学习，或者专家决策系统来解决问题。与人工智能不同，大数据的核心能力是要数清楚数。通过海量数据，我们能知道到底有多少资源，这些资源在哪个位置，很多时候当我们知道了这个事情本身，我们就可以让决策变得更精准。

以教育为例。我们可能需要知道的是，一个省到底有多少不同类型的教育设备，有多少教师，每个县短缺什么，有了这个数据我们就可以直接决策，并不需要很复杂的人工智能。所以这个边界我觉得大家有混淆，往往把云计算、大数据、人工智能甚至区块链等混为一谈。其实它们各自有各自的核心能力，如果没有明白这一点就容易犯一些错误。

当然最后一个误区，就是很多人过度恐惧大数据带来的坏处，比如说侵犯隐私。实际上真正因为隐私数据泄露而遭受巨大伤害的人屈指可数。又有很多人过度乐观，觉得好像未来一切皆会数据化。我觉得这些都是比较极端的想法。我们的确需要关注隐私问题，关注数据伦理问题。但不必谈虎色变，因为真正被老虎咬了的人并不多。我们应该以比较从容的心态去认知数据产业和数据科技的发展过程，其中既会有好的地方，也会有坏的地方，但是不能单方面地把好和坏放得特别大。

信风智库：作为智能时代的生产资料，大数据如何支撑人工智能与区块链这些新技术的发展？

周涛：你提到了支撑，其实本身就回答了这个问题，很大程度上大数据对人工智能是一个支撑关系。因为人工智能目前比较活跃的一个分支是深度学习，深度学习中的深度神经网络提出来已经几十年了，为什么在最近的10年这么火？很大程度上是因为ImageNet开放了大量标注了的图片数据，使我们可以用更多的参数在更大的空间中进行一些不需要知道任何因果关系的学习，这是需要海量数据作为支持的。所以在人工智能中有相当一部分依赖于标注好的高质量知识数据。

人工智能是生产工具，大数据是生产资料，而区块链是不一样的。区块链的建立与运行，完全不依赖大数据，自己就可以建一套东西。反过来，区块链可以反向赋能大数据，因为有了区块链这种加密工具，我们可以采集更可信的数据，对数据进行篡改与修饰会变得更加困难，甚至在技术上是不可能的。

当然区块链本身的成本也很高，所以也不是任何地方都要用。区块链不依赖大数据，只不过区块链上面没有数据就不会有价值。如果把大数据比作水，区块链就像水管一样，有了这条区块链水管，数据流通起来会更顺畅，否则会有大量的数据被随意修改，真实性无法保证。

信风智库：实际上我们看到不少大数据平台与系统还停留在信息化的阶段。大数据行业进一步发展，还需要解决哪些问题？

周涛：虽然在数据库、数据存储、计算能力以及计算资源的分配上，还有一些可以努力提升的地方，但对于大数据而言，其实现在制约我们发展的瓶颈并不多。面对大数据的下一步发展，更重要的是我们认知和管理思路的变化。比如政府如何用一种好的方式，解决数据开放共享过程中的安全隐私问题；如何在数据使用的过程中，去评估各个数据提供方提供的数据的价值，形成一个正向的、可感知的、可记录的反馈。这些问题都是需要我们解决的。

另外，要探索如何用一些新的办法使我们的系统建设不只是停留在信息化和业务功能的基础上。比如从数据质量的角度出发，数据监理与质量控制的方式，使业务系统能够产生高价值且可流通的数据。这里面还存在大量问题，甚至包括数据质量控制，当然也存在很多的技术问题。

所以，我们需要通过大量的努力打通这些理念和思路的瓶颈，这甚至比技术上的突破更加重要。众所周知，国家发文说数据要作为一种新型生产要素。但实际上数据产生后，还要通过多次流转流通，才能实现更大的价值，这条路我们仍还没有走通。

大数据需要"三高":高标准、高质量与高价值

信风智库:标准和质量是数据治理的两大核心。那么我们在实际应用过程中,应如何来设置数据的标准,如何来把控数据的质量?

周涛:第二个问题比较简单。对于质量的好坏,大家是有一些公认的标准的,比如数据的缺失率问题,举一个简单的例子,很多情况下我们的数据基本上就是一个大的表格,那么这张表格是不是能被填满,能填多满就是缺失率的问题。又比如,数据的实时性是不是很好,会不会有长时间的延迟;数据逻辑是不是自洽,数据中会不会出现两个人在同一所学校读书,但显示上学的城市不一样,这个逻辑就坏掉了。

然后就是数据会不会有一些很古怪的、异常的错误值。比如手机号码只有9位,那不就是显然有错误吗?还有包括数据的标注之间冲突多不多,标注可信度高不高,比如同一段方言录音,不同的标注人员写出来的文字差异很大,可信度就要打折扣。这些东西有一套整体的办法描述数据的质量。对于常见的、99%以上的数据,计算机程序能够自动对数据质量进行判断。所以这是可以做的事情,只不过现在应用还不广泛,因为

我们还不够重视数据质量。

第一个问题是很难的事情。数据标准看起来简单，因为我们时刻绷着一根弦，标准既是对行业的促进，也是对行业的束缚。一个行业在没有标准的时候，大家各行其是，百花齐放，百家争鸣，系统都不一样，将来没办法统一分析，显然需要一个标准来促成。但是一旦有了标准，就意味着在标准之外的很多东西要进来将变得非常困难，还有门槛，而且标准会给出一个边界，边界使不在标准之内但有可能更好的东西进不来。所以，标准同时也是行业的束缚。

我们在建立新的数据标准时一定要注意：

第一，我们可能要以需求出发，以终为始。我们需要弄清楚这个行业的需求是什么，做标准要解决什么问题，而不能闭门造车地建立标准。

第二，标准要有弹性，不能像以前一样建立那种硬化、固化的标准。现在我们建立一个标准的时候，可能要留出很多字段，比如原来可能是 12 位，那么现在我们可能要在 12 位的基础上，再留几位空白字段。将来，只要前面的基础字段符合标准，我们就可以用来做匿名化、做同一个 ID 的对标。以前标准写死了就这么多，将来有新的东西容纳不进去就很麻烦。所以建立数据标准我觉得更难，但是现在从总体上来说，在一些比较成熟的行业，特别是以前有数据标准的行业，推动起来还是比较顺利的。

信风智库：在规范化治理数据标准和数据质量的基础上，如何实现数据价值化和数据资产化？

周涛：数据要实现价值，它的核心是在于有没有场景，有场景的数据就能实现价值。举个例子，我们可以依靠技术大量获取电商平台上服装的价格和销量,从而很可能预测出当季会流行什么颜色和款式的服装。这很有趣但这对我们的价值很小,因为我们没有服装厂。而如果我们跟服装厂合作,这个数据就能指导生产。所以只有场景才能够使数据产生应用,产生价值。打造或找到适合数据应用的场景,是让数据发挥价值的最佳途径。

其实中国不缺场景,但还缺一件事情,就是知识付费。比如针对某个场景的大数据应用,不容易推广销售,因为是软件,但把这个大数据应用装到计算机上,再把计算机销售出去,就相对容易了,因为这是硬件。实际上在任何计算机上都可以安装这个应用。所以,怎样让大家愿意为数据价值变现的过程付费,也是一个问题。

数据资产化甚至资本化更为复杂。它意味着把数据写进会计表,写进企业的资产负债表,能做质押、抵押,甚至将来能作为股本金进行投资,这显然是一件更难的事。数据资产化需要有第三方的评估机构对数据的价格或者价值进行评估,这样才能促使数据资产化或资本化。而这种评估又面临很多问题。

首要问题是,我们可能需要一个持牌或者试点的评估机构来保证某一次交易,真正交易的是数据本身,而不是其他资源。数据交易所或数据交易中心能够进行数据产品登记和记录,并且认证某次数据交易的真实性与资产价值。

比如一家上市公司,在财报中显示花了500万元买了数据,证监会可能不认:怎么能证明这个数据值500万元?但是如果在你购买产品时有提前登记和评估,大数据交易中心认证了这笔数据交易的资产价值,那么数据资产才有可能得到更为广泛的认可。

最初,某种数据的价值只能根据需求方定价,即谁愿意花多少钱买就是什么价格。发展到一定程度，数据就可以有一个标准价格，比如租房平台的租房价格的数据，又比如一个给定精度为 10 平方千米的遥感数据，就可以有一个逐渐透明化的标准，进而逐渐形成一套评估定价的体系。

这就类似原来我们对知识产权的评估定价，最终还是体现在财报上，看知识产权产生了多少价值。利用这套方法我们可以对数据进行定价，有了持牌可信的评估机构，我们才能根据数据的所有权或使用权，以及交易记录来实现对数据的资产化和资本化。

数据要经历资源化、资产化和资本化三个递进阶段，这条路很有价值，同时也很难很长。我的观点是：道阻且长，行则将至！

数据要素市场化流通如何落地

信风智库：我们如何对数据进行定价？这个过程需要遵循哪些原则？

周涛：数据定价有几种方式。一种是数据本身已经有交易记录了，定价是比较简单的。我卖给你的数据，之前已经卖给过别人了。你愿意付100万元买这个数据，至少第三方评估机构可以评估你买的数据是值这个价的。所以你花100万元买数据写进你的财报是没有问题的。

标准化的数据产品也比较容易。因为这个行业有标准，我们有一套定价方式，比如通过遥感得到的大数据。

最难的其实就是企业有一套数据，但不知如何进行定价。以前企业通过生产一些产品能挣到很多钱，现在企业要把这些生产过程中产生的数据，就像研发投入或者其他资产一样进行定价，即说明这些数据是一项资产。

这方面可以参照知识产权在企业转化中的评估方式，根据企业财报对相关直接收益项目进行确认。但是这还需要一个系统化工程，就像我们曾经花了很多年来试点，才真正让市场认可知识产权评估这个事一样。

信风智库：跟土地、劳动力与资本一样，我们国家已经把数据列为七大市场要素之一。跟其他要素比较，数据要素有哪些特殊性？数据要素市场化流通需要哪些基础条件？

周涛：首先，数据具有生产要素的一些特性，比如它有稀缺性。但是它的稀缺性和土地的稀缺性不一样，土地就这么多，越用越少，而数据从某种意义上讲是越用越多，可以无限地复制。

但数据依然具有稀缺性，比如当我们想用遥感数据或者电商数据的时候，虽然那个数据可以无限复制，但如果我们没有，还是得花钱去买。所以，数据是越用越多，但又有稀缺性。稀缺性才是生产要素的一个关键，如果没有稀缺性，就不是一个经济流通中有价值的东西。

数据还有一个特性，就是在我们转移或交易数据的时候，往往转移或交易的不是所有权。其他要素比如资本，转移的实质往往是所有权。知识产权既可以转移所有权，又可以转移使用权，大多数情况下是以转移所有权进行投资，或者转移使用权进行授权再开发。数据在绝大部分情况下转移的是使用权，而不是所有权。

举个例子，怎么判断甲方卖给乙方的数据，乙方不再拿去卖呢？这暂时没有很好的办法。我们可以把数据放在一个介质中，比如一张光盘或一个 U 盘，你只能把 U 盘插在计算机上才能使用数据。但这种就不太好卖，因为用起来很麻烦。

为了避免我们卖出去的数据别人再拿去卖，我们就需要有一个交易中心来保障数据的交易。其中，数据交易中心除了能够记录交易行为，还可以保证二次转卖无法登记、不再有效或者不能交易。所以我们还需要建立各种各样的机制，保证在买卖数据使用权的同时，买家不能二次转卖使用权。

数据所有权还没有办法交易，因为所有权不明确。数据确权在短期内难以实现，而且意义也不大。比如某公司采集了很多的个人数据，这些数据的所有权到底是个人的，还是该公司的；或者政府委托一家公司提供一项免费服务，我同意了贡献数据，这些数据到底是受委托公司的，还是政府的，或者我的，这不好说。

短期内想把它说清楚，很可能会给大数据产业带来打击，而且作为个体拿到数据所有权后，也没有实际意义与价值。有些人想过是不是提交个人数据之后可以挣点小钱。但我认为，第一这对个人不一定有多大的好处，第二这对行业也没有明显的推动作用，所以这个事情也不一定能持久。

从这种意义上讲，数据作为生产要素，很有可能在很长一段时间内，在交易过程中处于所有权不明晰的状态。所以我们要保障的，不是交易物，而是交易这个行为，保障它的价值性与唯一性。

总地来说，这真的是一个新的事物，一个新的生产要素，所以它需要我们在法律甚至伦理方面进行更深的思考。

信风智库：数据交易中心是数据要素市场化流通的载体。最近，一些地方的数据交易中心开始运用诸如隐私计算等新技术，使数据可用而不可见，这会带来怎样的影响？

周涛：数据所有权和使用权在某些情况下是能分离的，有些情况下是不明确的，有些情况下甚至很难分离，那么，多方安全计算、联邦学习和隐私计算，实际上在一定程度上解决了部分问题。

比如几家银行提供数据共享，有可能互相付费，也有可能互相免费，然后通过多方安全计算或者联邦学习，各自得到想要的计算结果，同时又可以不让其他各方知道自己的客户数据。

然而，大部分情况下数据的提供方是单一的，只是提供方不想使用者拿走数据，这就不必采用联邦学习或多方安全计算的方式，而采用数据沙箱和隐私计算。你只能把程序放在我的数据沙箱里运行，你拿不走我的数据，只能得到计算结果。

这些新技术还是只能解决部分问题。仍然有很多情况是我要拿走你的数据才能计算。因为我可能还不确定用数据来算什么，我要根据你的数据来研究算什么，我也不想让你知道我算了什么，因为这很可能是商业机密，所以我必须拿走你的数据。

但是我拿走你的数据之后，你又不想我再卖，这是最难的问题，现在也没有特别好的解决方案。整体上看，我们依靠当前的技术逐步在为数据交易扫清障碍，比如至少 API 接口查询与调用目前就很轻松。但是要完全扫清障碍，我觉得还需要时间。

大数据如何赋能社会治理

信风智库：最近业界探讨的数据监理，主要解决什么问题，它对整个大数据产业发展有着怎样的意义和价值？

周涛：数据监理实际上是一种专业化的监理服务。目前随着大量信息化系统的建设，尽管业主方也要求这些信息化系统产生一些数据，但是没有专业机构来判断，系统是否按照业主方的要求产生质量达标的数据，数据是否符合相应的标准。这实际上是一件非常专业的事情，包括对数据本身逻辑和数值的一致性以及异常进行检测。

以前我们做信息化系统时，往往只对信息化本身的参数进行控制与检测，同时去判断它是否满足你的业务流程与需求。然而一种普遍的情况是，尽管它满足了业务需求，却不能满足数据需求。

举个例子，比如政府部门建设了一个帮助老百姓反映民情的服务系统，老百姓可以拍照片上传，政府部门也能处理。但如果我们想把所有涉及地面不平整或者裂缝的照片都收集起来，可能就做不到。很多系统只能一张一张地保存，那就意味着如果有10万张照片，就要保存10万次。从数据监理的角度来讲，这样的系统设计是没有办法按照一定需求顺利导出数据的。

还有一种普遍情况是，虽然我们可以对这些照片数据进行描述，形成一个高质量的标注数据，但是这些描述和照片在数据库中是分离的，所以我们无法链接它们形成有效的数据集。

当然，数据质量不好还有很多其他原因。比如系统设计可能为了美观好看，不管照片怎么拍，都强行地塞进显示框来显示。如果懂数据监理，就应该建立两个照片数据体系，一个适应显示框，另一个保持原始高清照片不变形。

为了保证形成高价值数据，我们需要形成独立于信息化指标和业务指标之外的第三套指标，然后由专业的公司和服务团队保障完成好第三套指标，这就是数据监理要做的事情。

信风智库：在疫情防控方面，大数据发挥了哪些作用？

周涛：整体上来说，大数据应用于疫情防控可分为三个层面：第一个层面是满足应急管理精准化。比如，疫情发生后，全国各地很快就有了健康码，跟三大通信运营商进行了数据共享，能够判断某个人在某地驻留了多少时间，以及大致的轨迹和位置。疫情初期有一个任务就是要找到所有从高中风险区流出的人，这使我们在防控疫情时，可以有的放矢。

后来，在信息化水平比较高的大城市，比如上海、杭州和成都这些地方，虽然出现了一些零星的反弹，但没有封城和封街。成都郫都区就出现过反弹，所有密切接触人群马上被找到并做核酸检测，没有影响成都的秩序，大家都正常上班工作。但是某些地方一发生疫情整个城都封了，全城所有人都要去做核酸检测。这是因为后者信息化还不够充分。这就是第二个层面，除了基本信息，还要掌握轨迹，并且还有技术办法使我们能够知道哪些人是密切接触者。

第三个层面就是进行整个传染病防控的社会仿真。比如我们和疫情较为严重的意大利合作，把城市划分成一个个 0.25×0.25 经纬度的格子，看格子里面有多少企业，多少学校，多少家庭。针对格子里的每一个人的真实统计数据，我们进行基于马尔科夫链蒙特卡洛方法的随机仿真。

这个时候我们就能判断每一步操作，每一个政策可能带来的影响。比如我们要去判断为什么意大利关闭学校并没有真正缓解疫情，发现他们采用的方式是，有一个学生被感染了，就封闭整个班，如果有三个学生被感染了，就封闭这所学校，但实际上病毒已经传播开了。后来我们建议按一定时间周期实施抗体检测，基于抗体检测来判断需要如何封闭学校，最终使学校的封闭时间缩短了，但疫情反而得到了更好的控制。

我们还在研究公共场所的疫情防控措施。基于每一个场所，比如一家超市，研究它可能有多少密切接触，然后决定是否封闭这个场所。这里面我们要看三个指标：第一个是对经济的影响；第二个是对民众的影响；第三个是能在多大程度上减少密切接触。

我们基于真实数据，通过大规模的仿真，建立了真实社会的一个数字孪生社会，进而可以测试各种各样的政策，去模拟它们对疫情防控的影响，从而精准地选择最好的政策。

信风智库：在社会治理现代化以及构建新型基层社会治理体系方面，大数据如何发挥作用？

周涛：大数据在社会治理方面的运用，最关键的是要利用大数据真正搞清楚有多少资源，这些资源在哪里，它们的使用效率怎么样。只有搞清楚了这些问题，才能真正找到问题的症结，而不是盲目地加大一些不必要的投资。

举个例子，我们只有获得了基于疫情防控的数据，才可能知道，一个城市到底有多少人，有多少人在写字楼里办公，新建的大量办公楼，到底有没有人使用，使用率究竟有多高。

又举个例子，比如在乡村进行公厕改革，就需要摸清楚一个村到底有多少公共厕所，每一个公共厕所有没有人用，有多少人用，使用频率如何。不能拍脑袋决定是否要再投入，或者不投入。

所以，用好大数据的精髓，第一步不是智能化的手段，而是真正利用这些数据去摸清楚基层有哪些资源，哪些是要再建设再投入的，哪些只需要把资源盘活即可。比如基层的图书馆、青少年活动中心、老年大学够不够用？这些就是数据能提供的帮助。

如何应对万物互联带来的数据大爆炸

信风智库：面对未来万物互联带来的数据大爆炸，在数据治理与安全方面，我们应该如何应对？

周涛：万物互联与传统互联网相比，最关键的差异不是数据来源，而是在于大宗数据是来源于一个可控的场景，还是从多个场景中合并而来。

举个例子，比如某家公司有很多互联网数据，且都来自这一家企业，我们只要管理好这一家企业，这些数据就能直接通过该公司下辖的平台为大家服务。这些服务既能够给企业回报，也能让消费者满意。

但是产业互联网则完全不一样。数据来自不同的工厂，这些工厂和工厂之间的管理主体也不一样。所以如果要用一个平台对这些数据进行采集或者管理，就会面对极高的复杂性，而且数据标准也不一样，也就不容易产生价值。

各家工厂的特点不一样，环境不一样，生产的东西不一样，如果一家一家地提供服务，其本质是做项目，没有普适性。如果把这些数据都拿过来，提供一个具有普适性的产品或平台让大家都来用，这样的难度很大，可能还得一家一家地谈，从而又变成了项目。

实际上，这是一个在数据分散且量很大的情况下怎样进行管理的问题。面向分散且零散的工厂，很难从过程进行管理。目前来看，更多是从结果来进行管理。管理方提供安全标准与技术服务，可以对工厂进行保护，如果因为工厂的问题，比如数据泄露，给别人带来了损害，就要从结果上对工厂进行处罚。这种管理方式，有点类似于在环境保护中管理大量排污企业，给标准给服务，最后根据排放结果来进行处罚或者奖励。

另外像无人驾驶、心脏起搏器或者机器人手臂等情况，我们要纳入特定场景进行管理。因为这个场景是有特别的安全隐患和需求的，一旦出现问题，可能会带来重大的人员伤亡。

信风智库：大数据时代下，我们如何在享受大数据带来的便利的同时，而不被大数据操控？

周涛：人为什么会被操控？这是因为在被操控的时候，他会觉得很快乐，精神层面获得了满足，从而把自己陷入一种信息茧房之中。比如在美国的社交网络中，我们会发现共和党人士关注的账号90%都是共和党人士，只有10%左右是跨党派关注。人们选择信息获取渠道的时候，会主动选择同类或者相近的人，因为在这样的社交网络中，人们会觉得快乐，说任何话，都有人应和。所以，我们不能把板子都打在大数据上，很多情况下是我们自己自愿被操控。

第二种情况，就是商家出于利益而进行操控，这种情况不一定是我们自愿的。比如我们在一些新闻网站或电商平台上，其实想发现一些新的东西，但商家却反复推送那些我们很感兴趣的或最有可能购买的东西。这种情况下，信息技术就变成一个凸透镜，让我们的视野变得越来越窄，甚至比之前更窄。这就是所谓的信息茧房。

如何打破信息茧房？目前我们在做一些研究，通过一个更好的信息导航技术，在不影响点击率的情况下，大幅度地提高多样化，让我们的视野越来越开阔。但是，这些新的技术在多大层面上能够得到认可，以及这些新的技术会不会成为行业主流，现在还不好说。

信风智库：业界常常提到"让数据做决策"，与之相反的是信息茧房的问题。我们有没有一个参照依据，来区分哪些决策是需要自己来做，哪些决策是可以交给大数据人工智能来做的？

周涛：我认为一切决策都要我们自己做。机器做出的行为，我们不叫决策，只能叫响应，按照一个固定程序进行响应。我们希望的，不是让数据做决策，而是让数据说话，让数据告诉我们最重要的信息，做决策的依然是我们。

举个特别简单的例子，当年爱迪生发明电灯泡，创立通用电气时，电气化还远未普及。通用电气早期的投资方做了调查问卷，他们向民众询问：如果有电这种东西，你觉得会有什么用？得到的结果是，大家都觉得没有什么用，因为很多民众对电完全没有概念，怎么可能知道有什么用。甚至就连爱迪生自己都觉得电的用途就只有电灯泡一个。

这个时候怎么决策？到底要不要投资通用电气公司，要不要推行电气化业务？这种情况下，只靠调查得来的数据是不行的。

数据可以帮我们做通俗的决策，而不能帮我们做创造性的、颠覆性的重大决策，更不能帮我们做需要冷暖自知的决策。比如大数据不能帮我们择偶，虽然我们好像可以有一套方法，让选择比较靠谱，但实际上两个人在交往后，是否情投意合，是不能由数据说话的。

　　我曾经多次举过一个例子：2006年世界杯时，德国队守门员莱曼通过数据统计，得知了对方球员射点球的角度偏好，防守住了对方球员的多个点球。但是，假设守门员通过数据统计，知道了对手射门角度的偏好，却又依据行业经验已经判断出对手必定要反其道而行之。这种情况下，如果还按照数据统计的结果来进行防守，那就没有道理了。

　　所以，数据给我们的，永远是客观的判断，判断不等于决策，最终的决策，还是要我们自己来做。

吴晓如：
智能化如何为生活添彩

吴晓如 博士，毕业于中国科学技术大学信息与通信工程专业。1999年作为联合创始人创办科大讯飞，现任科大讯飞总裁。2010年获国务院政府特殊津贴。曾多次主持、参加国家863重点项目和国家自然科学基金项目，取得了丰硕的科研成果。曾先后获得国家科技进步二等奖、信息产业重大技术发明奖等奖项。

科大讯飞股份有限公司成立于1999年，长期从事语音及语言、自然语言理解、机器学习推理及自主学习等核心技术研究并保持国际前沿技术水平；积极推动人工智能产品研发和行业应用落地，致力让机器"能听会说、能理解会思考"，用人工智能建设美好世界；两次荣获"国家科技进步奖"及中国信息产业自主创新荣誉"信息产业重大技术发明奖"，被任命为中文语音交互技术标准工作组组长单位，牵头制订中文语音技术标准；同时还获得了首批国家新一代人工智能开放创新平台、首个认知智能国家重点实验室、首个语音及语言信息处理国家工程实验室、国家863计划成果产业化基地等荣誉。

扫码观看
访谈精选视频

人工智能的系统创新如何解决社会发展的重大命题？人工智能的应用与发展，存在哪些认知误区与实践短板？在人工智能的应用过程中，我们应该以怎样的目标为导向？又应该遵循什么样的原则？人工智能企业如何规避技术有可能带来的负面影响，实现真正意义上的科技向善？

人工智能如何开启下一步创新

信风智库：科大讯飞目前正致力于"用人工智能的系统创新解决社会发展的重大命题"。我们如何理解"人工智能系统创新"？

吴晓如：这是我们一直在思考的事情，如何通过人工智能系统性的创新，解决一些社会发展的重大命题。这里有两个维度。第一个维度是我们以前的单点技术创新。比如语音识别技术，我们不断突破语音技术的难点，不断提高语音识别的准确度，同时也开发出了"讯飞听见"以及"录音笔"等一系列以语音技术为核心的衍生产品。在这个维度上，我们主要是站在技术的角度去考虑问题，攻关突破重要技术后，再来看能够解决哪些问题。

但是，解决单一的技术问题并非创新的全貌，技术创新也呈现出了更为总体化、系列化和结构化的新趋势。所以，第二个维度我们就倒过来了。从问题本身出发，面对这个问题，思考我们可以把哪些技术整合起来，形成一个针对具体问题的解决方案。

在我们深入诸如教育、医疗与交通等一些重要行业时，我们必须与这些行业的客户深入交流，站在行业的角度思考问题，以及怎么通过技术来解决这些问题，从而实现降本增效。

以教育行业为例。教育行业不会关心你的技术有多么创新，语音识别做得有多好，它们的实际痛点是老师和学生的负担很重，城乡教育资源分布不均等，这些都是社会发展中的重大命题。那么，我们就要去分析造成这些问题的因素有哪些，站在问题的角度去看技术。

比如怎么降低老师的负担，这里面就涉及大量繁重的备课和批改作业等工作。这绝不是仅仅通过一个语音识别或者图像识别技术就能解决的，背后可能还有一系列的技术运用问题。其中有一些是已有技术可以解决的，还有一些是技术上暂时不能解决的，这就需要我们进行技术攻关。

所以从这两个维度出发，我们提出了人工智能系统创新，就是要搞明白我们最后到底要解决一些什么问题，以及诱发这些问题的原因是什么，分别对应什么样的技术，从而统筹思考分步解决。

基于从单点技术到系统创新的变化，我们对人才的要求也发生了变化。以前我们以技术人员为主，而现在需要能够深入理解行业，把行业问题与人工智能技术结合起来，通过组合与集成不同技术来解决实际问题的复合型人才。这对人工智能技术进一步的应用发展来说，是一个非常重要的挑战。

信风智库： 回顾这些年来人工智能的应用与发展，我们存在着怎样的认知误区与实践短板，又该如何去弥补与提升？

吴晓如： 这一轮的人工智能应用发展是以深度学习作为理论背景的，主要脉络是：从理论变成技术再延伸出去解决社会问题。目前，人工智能技术不是包打天下的共性技术，它也只能解决一部分问题。虽然在某些特定的场景里面，人工智能已经非常强大，但要达到所谓的"智慧"水平还有很远的距离，我们应该理性地认识到这个差距。

2017年，国家专门在科大讯飞成立了认知智能国家重点实验室。认知智能有别于感知智能，比如语音识别和图像识别就属于感知智能的范畴，而认知智能更为高级，是人工智能理解语音、文字和图像想要表达的意思。就目前的理论基础与技术水平来看，我们想要打造一种强大的、通用的人工智能模型，实现认知智能，解决所有问题，基本上是不可能的，除非人工智能在理论上有重大突破。

当然，现有的人工智能是数据驱动的智能，与一些具体的应用场景结合以后，就可以化繁为简地解决一些实际问题了。比如在医疗中细分到诊断某一个特定病种的问题时，我们就可以组织医学专家，搜集整理针对这一特定病种的大量数据，同时把相关病理诊疗知识形成人工智能可以运用的知识图谱，从而针对这一特定病种解决问题。所以，即使没有重大的理论基础创新，我们也能在一个比较细小的领域里，把大量的数据、行业知识和人工智能算法结合起来，形成一个个针对性很强的解决方案。

目前的人工智能就处在这样一个结合特定场景解决特定问题的初级阶段。科大讯飞的语音识别技术也是如此，我们的目标不仅是识别任何地方的任何方言与语言，还要能结合特定的应用场景与行业领域来进行更精准的识别。

其实早在 10 年前，科大讯飞就已经在研究深度学习算法，我们还是非常期待在理论上出现重大的突破，哪怕没有在大框架上的突破，也希望现有理论能继续往前走。所以，科大讯飞现在和一些高校联合建立了实验室，特别是在最近一两年，我们和高校的数学系进行了深度合作，还联合一些从国外归来的数学家，期望在数理基础上走得更深更远。一旦在理论突破上有了一个明确的方向，人工智能就会进步得更快。

但是，理论突破是一个漫长的过程。在这个过程中，比较重要的还是把人工智能的一些技术与应用场景结合，把专家的行业知识进行技术转换，在特定的场景里解决问题。所以，我们必须理性地看清楚，目前的人工智能还不能解决所有问题或者超越人类智慧，但也不是什么问题都解决不了。

人工智能探索与应用有哪些路径

信风智库：在人工智能基础研究方面，科大讯飞做了哪些前沿探索？

吴晓如：经常有人会问，科大讯飞做的人工智能技术和其他公司的到底有什么不一样？其实，人工智能技术框架也是分层的。第一层是用现有的人工智能技术去研发产品，把自己定位成人工智能技术公司对外宣传；第二层是利用一些开源的人工智能技术框架，做一些语音或图像识别的产品，但他们只能在框架内去做一些事，会受到一定限制；第三层是整个语音或图像识别的深度学习算法体现，完全是由自己开发的，具有原创性，在这种情况下，我们可以在理论框架上进行一些调整，比如在模拟人脑的网络结构上做一些适配调整；再往下就是第四层，与数理基础研究进行结合。其实在一定程度上，深度学习就是从抽象、杂乱无序的信息中，一层一层地提取有效信息，从而提升这个模型或网络的信息提取能力。在数理基础上的研究，就是为了深度学习的某些算法能够突出原来的一些表征，解决深度学习的可解释性问题以及其他抽象问题。

科大讯飞就在做第四层的事情。我们正在联合数学和计算机领域的研究团队，比如和北京师范大学合作研究类脑认知等，通过不同学科之间的交叉碰撞产生一些成果。我们每年都能在一些重要的国际大赛上拿到好名次，这背后就是我们有自己原创的东西，把大家的基础研究力量汇聚到了一起。如果我们用开源框架，别人做到 100 分，我们最多也只能做到 90 分。

在前沿研究方面，国家层面也正在实施不少牵动性的项目。我们参与了一个科技部的项目——类人答题，就是让机器参加高考，看看它每年能考多少分，这就能体现参与机构人工智能算法的先进性。比如文理科的各种知识，如何把这些书本上的知识变成计算机能够理解的知识。整个人工智能的逻辑链条包括了对信息的基本描述、理解、分析、推理以及决策，这个链条就需要各种各样的研究团队加入进来，包括数学、计算机学、语言学和教育学等，大家整合在一起，一步步地往前推进研究。这个过程中，有计算机工程上的突破，有技术理论上的突破，有对文字、历史与地理这些专业知识进行本体描述方法上的突破，也有网络结构和数理建模上的突破。坦率地讲，这种突破不像计算机工程技术突破得那么快，但这些底层研究的进展，我们还是能够看见的，比如医疗上的病理诊断，教育上的作文批改等。

这一轮由深度学习理论引发的人工智能发展是从西方起源的。在基础研究上，我们与西方相比还是有差距的。不过，近年来国家在基础研究上的投入越来越大，我们也非常期待在大的基础研究上，大的数学、物理和神经学研究上，能够进展得更快。

信风智库： 在人工智能应用于人们生活方方面面的过程中，我们应该以怎样的目标为导向？又应该遵循什么样的原则？

吴晓如： 这要从两个角度来看。第一个角度是使用这个技术是否带来了价值，例如在教育上是不是让教学与学习效率得到了提升，是不是改善了教学质量，如果没有价值，肯定就没有人使用。第二个角度是这个技术是不是符合社会发展规律的长期需求以及政策要求。比如国家对教育培训领域进行治理，就是因为教育培训机构只解决了一些短期问题，孩子数学不好，就马上强化恶补，但强化完后，数学能力有没有提升也不确定。所以，技术应用一定要符合国家和社会的长期发展规律，不能做违背长期规律而挣短期快钱的事情。

这两个方面我们都要考虑。一方面技术带来的效率提升要让用户看得见摸得着，我们强调用统计数据来表明，比如我们帮助老师和学生具体提升了多少教学效率，减少了多少低效甚至无效的学习时间。

举个例子，现在孩子们的作业都是千人一面，考 95 分的孩子和考 55 分的孩子，做的都是相同的作业，这本来就不太合理。此前老师没有办法，一个班五六十个学生，不可能为每一个学生都单独布置一份作业。但现在，技术手段就可以帮助老师给每个学生画像，提供个性化的作业，比如成绩最好的学生在 50 道题里只用做 5 道题，老师再给他布置 5 道难度大一些的题。在这个程度上，我们就可以通过统计数据看到，这样的功能到底能为老师和学生节约多少时间，老师和学生到底愿不愿意使用。基于数据反馈，我们还要进一步优化算法功能与用户体验，使其越来越符合老师和学生的实际需求。这样一来，我们就在场景、数据和算法之间形成了一个良性循环。

另一方面是要与社会发展规律与国家政策合拍。比如教育不能"唯分数论"，不能为了短期提升分数而进行填鸭式的灌输。其实从长期来看，学生学习成绩的提升，背后一定是学习能力的提升，而学习能力又可以分解为语言表达能力、逻辑运算能力与推理能力等。对此，我们与北京师范大学和华东师范大学进行了深度合作，研究通过人工智能技术辅助教学，测试与提升学生的这些能力，进而实现短期内学习能力的明显提升，又符合教育发展的长期规律。

科大讯飞不仅要让机器能听会说、能理解会思考，还要让人工智能建设美好世界。这并不是空洞的口号。比如语音技术可以用于电话机器人，而电话机器人可以帮助疫情防控，也可以被用来拨打骚扰电话。那么，我们从技术的角度，就要采取过滤或屏蔽的手段，防止这类有损于社会的事情发生。

智能化为各行各业带来了哪些改善

信风智库：人工智能如何促进教育资源分配、因材施教以及素质教育评价？

吴晓如：这个问题很好。教育公平性问题是一个大家都很关注的问题。造成教育资源分布不均的原因有很多，比如学区房。技术只能解决一部分问题。

其中一个重要的着力点就是如何通过技术放大老师的能力。一个老师的教学面向两三个班级，能力再强也只能把两三个班级教好。随着在线教育的普及，老师的教学就可以不受限于物理空间，就可以影响更多的班级和学生。怎么影响呢？我们不仅可以让老师通过网络向学生单向传输知识，而且可以通过人工智能建立老师和学生的双向联系。首先，人工智能相当于给每个学生配备了一个辅助学习的机器人，可以评判当前学生的学习状态，供老师参考。其次，老师在完成教学后，可以面向一千个甚至一万个学生发送测试习题，人工智能可以辅助老师分析测试结果，进而优化教学内容和调整教学进度。最后还可以设置双师课堂，一个主讲课堂通过网络面向多个远端课堂，远端课堂现场设有辅导老师，既可以和主讲老师进行交流互动，又能现场关注学生，让学生更好地进入学习状态。尤其是在语言学习上，不少乡村地区可能没有英语老师，我们就可以用人工智能技术进行语音合成，实现与真人英语老师差不多的语言教学与评价。我们还可以用人工智能技术为作文评分，做到与真人老师不相上下，从而辅助乡村老师开设相关课程。这些智能化的辅助手段，能够在一定程度上让更多的学校尤其是乡村地区的师生获得优质的教育资源。

针对第二个因材施教的问题，我们首先是把老师和学生的教学过程进行数据化，比如作业情况、考试成绩以及师生互动等，一些学校还让学生戴上手环，监测体质与运动等数据。这些数据积累了很多之后，就可以通过人工智能分析处理，不断地给学生画像，随着数据越来越多，时间越来越长，画像就越来越深，人工智能就可以给学生提供更加个性化的学习方案，减少学生低效和无效的学习时间。节省下来的时间，就可以用于综合能力和创新能力的培养。如果我们仍然按照20年前的教育模式，下一代孩子不见得能够适应未来社会。所以一方面我们要在现有的学业体系里让学生更加精准地学习；另一方面提供一些创新学习的手段，比如我们正与一些学校合作开展人工智能和语言方面的创新教育。

第三问题是一个比较大的问题。尽管现在还没有一种评价手段能代替高考，但国家也在鼓励打破"唯分数论"，提到了需要对学校、老师和孩子搭建新的评价体系。对于学校和老师，不仅仅看高考考了多个清华北大，对于学生除了高考，还要有选拔人才的其他通道。那么，怎么去打破"一考定终身"呢？那就要在日常学习表现中，能够对学生形成一个更为立体的评价机制。这里面包含了两个方面，一方面是评价手段高效准确，大数据人工智能技术要和现在的教育体系相吻合；另一方面是评价过程要更加公开透明，要更有社会公信力，比如利用区块链不可更改的特性。这样就能形成一种以技术为关键支撑的先进评价体系，我们已经在进行相关的探索与尝试了。当这种评价体系的公信力越来越高、越来越准的时候，就有可能使现在高考选拔人才的制度发生变化。

信风智库： 人工智能如何促进医疗资源分配，如何提升诊疗效率，以及如何助力疫情防控？

吴晓如： 在医疗领域，我们在2017年做了一件令自己信心倍增的事情。当时科大讯飞的一个智慧医疗机器人参加了全国执业医师资格考试，考试结果超过了96%的考生。所以我们觉得让一个机器人去理解医学的本体知识，可以达到一个比较高的水平。

当然，医疗领域还有很多的问题需要解决。第一个就是分级诊疗的问题。病人都一股脑地到三甲医院去治病，造成了两个极端——三甲医院里拥挤不堪，一般医院却没什么病人。我们希望能够强化基层，通过一个技术手段，辅助基层医生诊断治疗更加准确有效。这样一来，老百姓遇到一些小病或常见病时，就没有必要去三甲医院，在社区医院就能获得很好的医疗服务。三甲医院只讲解3分钟，社区医院则会详细地讲解10~20分钟。我们把这种技术叫作智慧辅诊，目前已经实现每天30多万次辅助诊疗，一定程度上提升了基层医院的诊疗能力，在基层医院解决了小病和常见病。

第二个问题我们以慢性病诊疗为例。我国的慢性病人群非常庞大，他们在日常生活中需要药物治疗，也非常希望得到医生的关心。但显然我们的医生无法兼顾这么多病人。所以我们与医院合作开发了一个慢性病管理与医护关心的智能化应用系统。病人在日常生活中，可以通过App或电话的方式问诊。当前身体反应如何？吃了药以后会不会过敏？恢复情况怎么样？饮食和用药应该怎么调整？我们将病人向医生沟通询问的常见问题进行模式化，再通过智能化的方式让机器自动回答这些问题。以前可能有100个问题需要医生来回答，现在机器就可以帮忙回答95个常见问题了。

第三个是涉及国家传染病防控的一些问题。原来的传染病防控是基层医院发现疫情后一级一级地往上报，疾控中心再进行分析。现在我们在很多社区里都设置了传染病防控和预警机制，如果几个社区同时发现了一种传染病，将多点触发瞬时启动预警机制，疾控中心可以立即拿到预警报告。同时，基于这几个社区的人口库，立即使用疫情自动排查系统，可以电话联系到这几个社区里的每个人。这样一来，传染病预警速度和排查效率就能大幅提升。

信风智库：汽车是人工智能应用最深的场景之一，科大讯飞是如何为汽车赋能的？

吴晓如：人工智能是最能够激发想象力的技术，同时汽车是集大成的技术应用平台，所以人工智能和汽车结合起来，就可以创造出无限的想象空间。

过去几年，科大讯飞在汽车领域中的应用越来越深入。我们与上汽、吉利和长安等很多汽车厂商展开合作，语音交互系统已经应用在了1000多种车型的2300多万辆车上。如今，配备语音交互功能模块的汽车越来越多，我们粗略统计过，每辆新出厂的汽车中，就有一辆配备了语音交互系统，其中每10辆配备了语音交互系统的汽车中，又有7辆使用了由我们提供的技术。

我们可以看到，诸如自动驾驶、语音识别、图像识别、认知智能等越来越多的人工智能技术应用到汽车上，会使未来的汽车变得更加聪明，也更加贴心。

如何体现汽车的聪明贴心呢？现在的很多汽车，用户按一个键就可以发出各种各样的语音指令，这种应用虽然已经体现了人工智能的聪明性，但是还不够贴心。未来不用按一个键，汽车就可以随时判断当前驾驶者到底是在跟汽车说话，还是在和旁边的乘客说话。驾驶员在驾驶的时候，可以从容地把汽车看作自己的朋友，让它提供各种各样的服务，以后汽车将更加聪明、更加仔细、更加准确地接受和执行各种各样的命令。

同时，未来的汽车不仅能够"察言"，还能够"观色"。汽车未来就是一个多媒体、多通道、多样式的人机交互场景。通过在车内安装摄像头，无论是对坐在前排的驾驶员，还是对坐在后排的乘客，汽车都能非常清楚地给予各种各样的服务。对于消费者来说，汽车和手机不一样，手机一两年换一次，而汽车是耐用型消费品，很长一段时间里这个家庭都会和这辆车打交道。所以在应用过程中，汽车就能通过"察言观色"，不断理解家庭成员的偏好，了解每个人的喜好，从而为每个人提供越来越个性化的服务。

汽车也是时尚舒适的移动空间。在最新技术进展中，通过人工智能技术和其他技术应用的融合，汽车可以实现自动降噪，高速行车环境下通话会更加清晰。针对车内的音响效果，原来调音需要很长时间的积累，需要有多年经验的专家深度参与，现在这些专家和人工智能技术配合起来，就可以更加高效地将声音效果调至最佳。大家很快就可以在十几万元的车上享受到百万元级别的车载音响效果。

汽车更是安全可靠的移动空间。对汽车厂商来说，不仅要把车研发好和造出来，而且要在整个使用过程中，给用户提供更加安全可靠的服务。就像劳斯莱斯公司通过网络和算法远程监控飞机发动机的运转情况一样，汽车厂商也可以建立汽车与车厂的连接，提供从意向购买到销售流程，再到使用过程的全生命周期服务。汽车的静态与动态数据经过脱敏处理后，就可以供给汽车厂商实时掌握每辆车的运行情况，通过算法进行及时的故障风险预警并及时给出应急处理方案，从而保障每辆汽车安全可靠。

从人工智能的角度来看，未来的汽车将是一个集合了各种视觉、听觉、对话系统的人工智能平台。我们可以通过网络把每一辆车与车上每一位用户的数据经过脱敏处理后，存储在后台的用户数据平台中。这样就可以进一步通过人工智能、大数据、云计算等先进网络的融合，让移动空间变得更加智能化，赋予汽车更加聪明的大脑，为每个用户提供量身定制的服务和美好的出行体验。

人工智能如何肩负起社会责任

信风智库：人工智能如何更好地服务老年人，如何帮助老年人更好地拥抱智能化？

吴晓如：我们很早以前就在关注这个问题了。但是面向老年人群的一些智慧化应用还是不容易做。对此我们考虑和设计了三个维度的产品，有些已经做出来了，有些还在进一步的开发中。

第一个维度是，如何改造现有的智慧化服务，使老年人更容易使用。年轻人可以非常娴熟地使用各种网站和 App，但老年人学习起来难度较大。政府层面也非常重视这个问题，对政务服务与公共服务进行适老化改造，让老年人办理证件和申请补助更加方便。比如语音就是一种人与人之间最简便、最自然的交流方式，老年人可以很容易地用语音去查询各种服务、询问各种问题。我们与一些地方政府合作，对现有的政府热线和 App 进行智能化改造，让老年人更容易地融入现在的数字化生活。

第二个维度是对老年人的健康与安全进行监测和辅助。现在很多社区里，老年人独自在家而子女都不在身边，社区人员又不可能天天敲门关心询问，可能会出现老年人生病甚至去世了,很长时间都没人知道的情况。所以我们做了一些智能感知系统,比如在老年人家门外安装一个摄像头,以及把家里的水电气与管理后台连接起来。如果发现老人连续几天都没有出门以及家里连续几天都没有用气,系统就会自动拨打子女与社区工作人员的电话,告知他们去家里看看情况。

第三个维度是在一定程度上缓解老年人生活孤独的问题。独居老人不能天天和子女交流沟通,对此我们正在探索一些与老年人进行情感交流的服务和产品,这里面涉及很多人性与伦理方面的问题,我们还在思考。而且如何持续提供一些老年人喜闻乐见、愿意长期使用的服务和产品,也是一个难题。现在大家都在关注"科技适老",我们认为在老年社会与银发经济中,人工智能和机器人还能做更多的事情。现在在一些酒店,顾客打电话请人送一些牙刷或水果之类的东西,过一会儿门铃一响打开房门,就可以看见一个机器人拿着东西站在门口。类似这样的应用,是可以用到老年人社区的便民生活服务当中的。所以我们将人工智能技术开放出来,通过讯飞开放平台,帮助更多的合作伙伴与技术团队开发出各种各样的适老化智能应用。

信风智库:作为人工智能企业,如何规避技术有可能带来的负面影响,实现真正意义上的科技向善?

吴晓如:其实这是一个企业的态度问题。一种是凡是可以挣钱的事情,只要不违法我都干;另一种是希望把我们的技术用于社会阳光积极的一面,但如果有一些负面影响,哪怕法律不禁止,我们也绝对不干。这是一个企业的基本态度。

科大讯飞树立了"让人工智能建设美好世界"的企业愿景,而且不断地在内部宣传贯彻,从技术研发到产品服务,都树立了这样的态度。不做无益于社会的事情,哪怕很赚钱。

在这个基本态度影响下,我们建立了一套内部做事必须遵循的行为准则和方法。例如在数据隐私上,我们严格要求数据的"采存用"绝对不和个人身份相关联,从源头开始对隐私保护严格把关。又例如我们在教育领域进行数据建模与建立算法时,会聘请教育领域的专家学者,从教育专业和规律的角度告知我们,在对学生进行个人画像和提供个性化学习方案时,应该注重哪些方面,而哪些方面是绝对不能碰的,碰了反而会加重学生的焦虑与学习负担。

这些都是我们企业基本态度的体现。从数据安全和数据隐私,到每个员工的行为准则,再到最后的产品与服务,都是企业文化与基本态度的始终贯穿和匹配落地,而不是仅把口号挂在墙上。

肖风：
区块链如何构建新型生产关系

肖风　中国万向控股有限公司副董事长兼执行董事，上海万向区块链股份公司董事长兼总经理，万向区块链实验室创始人，南开大学经济学博士，有超过20年的证券从业经历和资产管理经验。历任深圳康佳电子集团股份有限公司董事会秘书兼股证委员会主任，中国人民银行深圳经济特区分行证券管理处科长、副处长，深圳市证券管理办公室副处长、处长，证管办副主任，博时基金管理有限公司总裁。

万向区块链致力于构建一个充满活力的全球化区块链生态圈，从技术、资金、资源等方面全力推动中国区块链行业的发展和落地。目前，万向区块链在区块链领域累计投入超过10亿元人民币，万向区块链生态在全球投资的优质区块链项目已超过150个。

万向区块链联合生态合作伙伴打造了以隐私计算为特色的新一代联盟区块链平台——PlatONE，通过分布式认知技术创新形成数字经济的可信基础设施。在产品应用方面，万向区块链聚焦供应链金融、智慧城市基础设施、可再生能源、生物资产等重点行业，以技术支持实体经济发展，同时被列为上海市高新技术企业、中国区块链技术和产业发展论坛副理事长单位、中国互联网金融协会会员与中国支付清算协会会员。

扫码观看
访谈精选视频

如何看待区块链所代表的新型生产关系？如何理解以区块链为代表的价值互联网？区块链中的分布式技术可以应用到哪些领域？通证经济模式将如何影响我们的生产生活？区块链如何在数据可信与安全方面发挥作用？中国发行的主权数字货币将带来怎样的影响与意义？

新型生产关系新在哪里

信风智库：对智能时代，大家形成了一个共识，人工智能是生产力，大数据是生产资料，区块链代表了新型生产关系。在您看来，这种新型生产关系究竟新在哪里？

肖风：区块链可以从两个层次来看，一是从纯粹的技术层次看，它本身就是生产力；二是从经济学的层次看，区块链技术对重构生产关系确实会起到一个非常新且重大的作用。

区块链非常巧妙地将博弈论引入进来，尤其是涉及了一种基于博弈论的机制设计，比如分布式账本与智能合约等，这是经济学里很重要的一个课题。

所以在这个意义上，我们可以说区块链改变了生产关系，这是很重要的一个改变。其实从加密数字货币就可以看出，这个世界上有成千上万的人可以设计出加密数字货币这样的电子货币系统。但如果不是基于区块链来做这件事情，就没有可信度，因为基于互联网的电子货币系统一直都有。只有区块链技术才能确保电子货币的确定性与唯一性。属于谁不属于谁，区块链网络就能确认了，不再需要第三方来确认。

一串字符存放在一台计算机上，你怎么说这个东西就一定值钱呢？怎么让大家认可它的价值呢？你不可能把货币发送给一万个人。但区块链的分布式账本可以像发送电子邮件一样，边际成本为零地与一万个人同步这串字符的价值。

可以说，区块链重组了很多东西，重组了价值流通，还重组了组织。比如公司为什么会存在？因为我们需要通过一个组织机构来更好地获得信任，更好地从事经济活动。

诺贝尔经济学奖获得者科斯先生就曾指出，公司这种形式能够降低交易成本。但如果通过区块链这么一套数字化的技术系统，就能够把交易成本变为零，这个时候公司的存在就不是必然了。因为即使不通过公司，我们仍然可以把交易成本降到非常低，低到经济交换没有摩擦，区块链技术就是从这个角度带来了生产关系上的重构。

当我们面对数字经济大潮的时候，区块链技术为我们带来了一种新的组织方式，这是非常适合数字化的，也是非常适合数字经济的。

信风智库：有一种说法认为，传统意义上的互联网是信息互联网，而区块链是价值互联网。那么，这个价值到底体现在哪些方面？

肖风：网络其实经历了三个阶段，最早是从发明无线电开始，人类社会首先架构了通信网络，电报与电话随之出现。20世纪下半叶，互联网被发明了出来。

互联网带来了什么？带来了信息网络。我们不仅可以通信，还可以处理数据、处理邮件等。信息网络给人类社会带来了巨大变化，不管是阿里、腾讯，还是谷歌、微软，它们都诞生于信息网络时代。

我们现在正从通信网络、信息网络迈向第三个阶段——价值网络。价值网络并不只是金钱的价值，而是基于价值网络可以帮助人们更好地管理资源。在信息网络时代，这些资源管理起来是非常昂贵的。

比如电商，我有货，你有钱，你要买我的货，我把货给你，我们双方在网上是陌生的关系。要保障交易完成，我们就需要第三方系统来担保。没有第三方担保系统，电商一定是做不成的，所以2003年淘宝网上线，2004年支付宝就出来了。

到了价值网络时代，比如在区块链上，价值交换与信息交互都不需要第三方进行担保。那么是用什么来担保呢？我们用区块链来做担保，主要是依靠技术系统与数学规则，这就是价值网络的核心。

我们前面讲过，所有数字的活动都是零摩擦的，边际成本是零。用零摩擦的一套系统来担保，显然比一个庞大的第三方网络系统进行担保的成本要低得多。这就是时代一点点地往前走，走到今天我们把区块链叫作价值网络。实际上，你也可以用另外一个词汇，比如说资源管理网络。

信风智库：如您所说，信息互联网催生了BAT等互联网巨头，那么价值互联网将催生怎样的新商业模式？

肖风：从信息互联网走向价值互联网，或者说资源管理型互联网的时候，商业模式与前面的信息互联网时代就有了巨大的差别。有一个很好的案例可以让大家理解什么是互联网商业，什么是区块链商业。

在美国有一家数字货币交易所，叫 Coinbase。它于 2012 年成立，2021 年 4 月在纳斯达克上市，市值一度超过 1000 亿美元。1000 亿美元是什么概念？纽约交易所和纳斯达克两个交易所的总市值加在一起大概就是 1000 亿美元。

Coinbase 有 1400 个员工在运行这样一个网络。但同时在区块链上有一个一模一样的，也是从事数字货币交易的平台系统，叫 Uniswap。Uniswap 是完全基于区块链建立起来的，它只有 11 个人。它一天 24 小时的交易量最多是多少？是 70 多亿美元。这已经达到甚至超过了 Coinbase 的交易量。

但你知道这 11 个人为 Uniswap 这个系统一共写了多少行代码吗？ 500 行。11 个人 500 行代码，这就是区块链商业。Coinbase 是用互联网的方法建立起来的，所以它需要 1400 人，需要成千上万行代码，需要有一个公司从头到尾负责信息安全，负责系统运营，负责管理客户资产，负责撮合客户交易，负责清算结算。

而 Uniswap 基于区块链，11 个人 500 行代码就干了同样的事情。难道 Coinbase 干的这么多事情，都不用干了吗？不是的。

在互联网上面，你需要全程构建一个方方面面的系统，你要负所有的责任。但区块链把所有的东西都提供了，比如信任，我们把区块链叫作信任的机器。因为两个人达成交易，首先肯定得互相信得过。产生信任的过程、安全以及钱包的功能，都由链帮你完成了，你把资产放到基于区块链的钱包里就行了，区块链天生就是一个清算网络，清算也帮你干完了。所以你要干的事情很简单，只需要把交易的规则用 500 行代码写清楚就可以了。

这就是互联网商业和区块链商业的巨大不同，这种不同不是 10 倍能效的提升，就 Coinbase 和 Uniswap 这两个案例来说，这是 100 倍能效的提升。我们可以预计，将来很多基于区块链的应用，其创新能力会有多么巨大。

万物互联需要去中心化

信风智库：分布式是区块链很大的一个特点。它不仅是一项技术，更是一种理念和一种模式。随着数字经济的发展，分布式可以应用到哪些领域？

肖风：分布式商业或者说分布式经济，都源于区块链最底层的一个分布式网络，与之相对的是传统意义上的中心化网络。当网络变得越来越复杂，需要处理的东西越来越多，需要涉及的面越来越广之后，中心化网络就不能承载了。

所以那些大型的互联网公司在搭建底层架构的时候，都逐渐走向了分布式网络与分布式存储。这是因为网络世界越来越复杂，你无法用一个中心化的技术架构来承载这么复杂的系统以及这么海量的数据。

大家有没有想过，如果 5G 与物联网技术发展起来后，人与设备、设备与设备都连接在网上，将会发生怎样的情况？根据 IBM 公司的统计，到 2045 年，全球将有 1000 亿台设备联网。联网并不只是把网络连接起来，而是要通过传感器去收集这 1000 亿台设备的数据，这些数据每分每秒都在产生，而且还要互相传输，中心化的网络承受得了吗？承受不了。所以必然走向去中心化的分布式网络。

举一个很简单的例子，就是现在热门的智能汽车。现在有两条路线，一条是车自身的智能化，另一条是我国正在主导的 V2X 车路协同。那么，车路协同的时候，我们要将所有数据上传到一个数据中心，让它处理完了之后再下载回来吗？肯定不能。我们必须在这辆车通过这条路的时候，端到端地直接发生数据交换，立即做出正确的决定，是要刹车、减速、左转还是右转。这就是一个完全的去中心化网络。

当世界大到 1000 亿台设备都在互相通信的时候，设备之间 80% 以上的通信，都不大可能也不应该被传送到一个统一的数据中心。因为等它算完后，将结果告诉终端再下指示，可能已经发生无数个车祸了。

在这种情况下，我们目前理解的分布式网络，已经不足以支撑这么复杂的一个系统了。所以它只能去中心化，就让这辆车和路边的设备直接交换数据，随即做出决定，然后再往前开 10 米，又有另外一台路边的设备告诉这辆车，有个老太太要穿过马路了。这辆车必须当场立刻做出决定。

从网络技术角度来说，从中心化走向分布式，再走向去中心化，这是最底层的一个技术架构变化。因为整个世界越来越数字化，越来越网络化，越来越线上化之后，必须要有新的底层技术架构与之匹配且提供支撑，而区块链恰恰就是一个可以帮助我们建立去中心网络的工具。

去中心化并不是说要去掉监管，更不是说不需要法律约束。而是智能时代的世界，需要这样一个去中心化的网络来管理，这里面当然有规则。比如车路协同系统，国家肯定要出台相关法律法规，包括智能汽车如何处理路上的突发情况等。

大家想想未来的物联网，所有的设备有这么多的传感器，在这个基础上，会催生分布式经济与分布式商业，当然也会是去中心化的。因为如果最底层的技术架构与资源管理模式都是去中心化的，那么它的经济、它的商业、它的金融也一定会变成去中心化的。

信风智库：去中心化是解决复杂性问题的手段而非目的，我们不能一概而论。那么，在实际应用过程中，去中心化是否也要分领域、分场景以及分阶段？

肖风：去中心化不能解决所有问题。这个世界也并不是所有事物都需要去中心化。就好像我们经历过农业革命、工业革命与信息革命，难道现在就没有农业了吗？还是有的。

从效率的角度来看，有时候越中心化越有效率，越去中心化越没有效率。

比如用现有的区块链来支付或者汇兑，就是一件没有效率而且费用很高的事情。对平台来说，要做到每秒50万笔交易确认，就只能用支付宝这样的中心化架构。

因为在支付宝上成交是由一个人来确认的，效率也最高。而在区块链上则需要几百个、几千个乃至几万个人和节点，来进行全网确认，这显然没有效率。

这个世界上还有很多事情是需要效率的，但有些事情也需要公平。有些事情可能需要大家来见证并达成共识，但也并不是所有事情都需要共识。

比如中国人民银行在设计和测试央行数字货币时说得很清楚，人民币数字货币目前并不依赖区块链技术。这为什么还不一样呢？

中国人民银行是法定授权可以发行货币的部门，背后是国家以信用背书，不再需要区块链来提供担保。但在发行之后的流通环节，央行数字货币或许也需要运用区块链技术。

通证经济将如何影响我们生产生活

信风智库：您之前谈到区块链是一种配置资源的方式。从经济学的角度来看，区块链代表的通证经济模式将如何影响我们的生产生活？

肖风：你问得很好。其实当初把 Token 翻译成通证，我认为不是太贴切。区块链发展至今，最近催生出了很热门的 NTF（非同质化代币），这才是一个非常纯粹的通证。

我们发现今天的区块链，作为资源管理与价值交换的工具，已经非常成熟了。它包括了三个种类。

第一类与加密数字货币一样，是基于区块链发行的数字货币，英文叫作 Coin。它和央行发行的国家主权数字货币不一样，是原生的数字货币，具有货币的部分属性。

第二类英文叫 Token，在 NFT 分离出去后，Token 就成了纯粹的基于区块链的代币。代币首先必须标准化，同一个事物的代币不能不同。它必须是一个很标准的价值衡量工具，代表某一项权利。代币是基于加密数字货币发散出来的，相对于第一类数字货币，代币更像数字资产。

第三类 NFT 就是一个纯粹的通证。它是非同质化的，独一无二的。NFT 用来做什么呢？比如用来确权，这个东西是你的，不是我的。我们把一幅画对应一个 NFT，谁有这个 NFT，这幅画就归谁。

除了确权，NFT 还用来授权。比如办公大楼的门禁管理，每个人都有一个通行证，通行证一定是非同质化的。你是访客，就只能用一次，我是在这个楼里办公的，我始终都可以使用。人们在不同的楼层工作，就只能进入对应的楼层，不能到别的楼层。这就是通证，也可以把它叫作令牌。

Coin、Token 和 NTF，数字货币、数字代币和数字通证，区块链发展至今，已经形成了比较完善的三大工具，可以帮助我们去架构基于区块链的分布式商业或者分布式经济。

那么回到刚才的问题。基于区块链的通证经济到底改变了什么？或者带来了怎样的巨大改变？

我们举一个例子，比如两个创业项目。如果你是以互联网的方式创业，你就需要写一个商业计划书去找风险投资。如果风险投资相信你的商业计划书和团队，那么他们就要跟你谈一个核心问题，估值是多少？如果是 1000 万元人民币，我先给你 100 万元的天使投资，过了半年以后，你的商业模式证明成功了，风险投资再给你 3000 万元的估值和新一轮的投资金额。

而区块链上的创业逻辑不是这样的。它是先把项目建立起来，然后所有参与者来分享 Token。最初的参与者并不知道拿到的 Token 会不会值钱，有没有价值。

实际上 Token 有没有价值取决于参与者有多少，是不是都积极参与，是不是都把聪明才智带到这个项目里面来了。如果参与的人越来越多，大家都很积极地参与，同时都为这个项目贡献了很多力量，那么这个 Token 就值钱了。这本身就是一个巨大的革新。

这就像几十年以前，农村生产队是记工分的。一个青壮劳力出一天工记 10 个工分。我那时是小学生，农忙的时候帮忙一天记 1 个工分。工分值多少钱是不知道的，只有到了年底的时候，生产队的会计算账，算算我们生产队这一年卖了多少稻谷，卖了几头猪几头牛，收入了多少钱。

比如收入 10 万块钱，同时整个生产队这一年积累了 1 万个工分。那么生产队就开大会宣布，今年 1 个工分值 10 块钱。

与互联网创业的预先估值不同，你在区块链上做一件事情，我给了你一个 Token，我给你的时候，它到底值多少钱？其实是不确定的。只有干了事情之后才知道。所以这从根本上改变了创业，把创业的门槛降到了地板上，降到了最低。

再以 Uniswap 为例，你写 500 行代码，你需要多少创业资金？根本不需要天使投资。现在市值是多少？接近 300 亿美元。这 300 亿美元怎么来的？是大家带来的。500 行代码只是写完一个交易合约而已，然后所有的参与者都用这个合约做交易，因此这个项目就值钱了。

所以我经常开玩笑说，像 Uniswap 这样基于区块链的创业项目，只有一个条件，如果父母还管你饭吃，你就可以创业了。你待在自己家里，父母每天给你做三顿饭，你就写 500 行代码，如果没人用，失败了也没有关系，反正就 500 行代码，可以再写一个，又不是什么大不了的事情？

信风智库：通证经济模式如何连接与激励利益相关体？

肖风：这就是 Token 带来的改变，把传统意义上的股东变成了更广义的利益相关者。它不只是对股东友好。20 世纪 70 年代创造出来的词汇"股东利益最大化"，在区块链时代完全失效了。

强调股东利益最大化，上市公司 CEO 的行为就会越来越短期化。因为每个月要出报表，每个季度要对外公布盈利与收入，还必须体现出增长。这就是股东利益最大化。

在互联网时代，实际上已经有很多企业家宣称不做股东利益最大化的事情，而是把员工放在第一位，最后才是股东。过分强调股东利益最大化，对公司的长期价值是没有帮助的，还有可能造成损害。

到了区块链时代，大部分项目没有股东，但是有大量持有 Token 的利益相关者，这个项目做好了，他们就都有利益上的激励，所以大家都有参与投入的意愿。这就是通证经济带来的变革。

当然，任何事物都有两面性。一些不法机构打着区块链的幌子，进行非法融资、非法传销与金融诈骗，既扰乱了社会经济秩序，又阻碍了区块链本身的应用发展。对此，首先我们应该从政策法规层面完善监管体系，加大打击力度；其次区块链从业者应该严格遵守法律法规，正确利用区块链技术，开发基于真实场景的创新应用；最后社会公众也应该依据法律法规理性判别，谨慎对待，增强金融风险防范意识。

信风智库：区块链领域的从业者，如何寻找适合区块链发挥价值的应用场景，这些场景有哪些共性特点？

肖风：大家首先要有一个最核心的观念，不能只把区块链看作是一种工具或者一项技术。真的要用区块链来创业，就要想清楚究竟用区块链来改变什么。因为区块链不仅是一项技术工具，更是一种经济模式。这就需要我们在对区块链进行创新应用时，设计好一个完善的博弈机制与商业模式。所以，真正意义上的区块链创新应用是一件很复杂的事情。

我见过很多在互联网做得非常好的人，他们要理解区块链，却成了一件很费工夫的事情。因为区块链的逻辑完全倒过来了，或者说是另外一套逻辑。他们在互联网领域越成功，跳到区块链领域就越困难。

所以，我们一定要明白，区块链是一个综合性的东西，自己没有发明任何新的技术，可是它把过去几十年的多种技术融合在了一起，做了一个融合创新，形成了一个非常精妙的区块链系统。

同时，我们尤其需要注意，一定不要为了区块链而区块链，不要为了去中心化而去中心化。如果用其他方法能够更好地解决问题，比如用中心化的方法能够更好地确认交易，就一定不要用区块链。

一定要找到一个场景，只能用区块链或者去中心化的方式，才能最好地解决问题。只有这样才有可能会成功。比如 Uniswap 只用 11 个人 500 行代码，就能做到每天 70 亿美元的交易量。还有比它更好的吗？还有比它更轻、成本更低、效率更高的吗？目前找不着了，它当然会成功。

区块链天然匹配数据要素市场化流通

信风智库：区块链的非对称加密、分布式账本与共识机制等特性，如何在数据可信与安全方面发挥作用？

肖风：这里有两个问题。首先第一个问题是收集来的数据到底可不可靠，可不可信？我们现在的模式是由一家中心化的机构或者用中心化的方式来保证。不管是大数据局也好，某个公司也好，收集数据的真实性与可靠性，都是靠这个方式来保证。

但是在将来，1000亿台设备联网，区块链是不是保证数据真实可靠的一个最佳方法呢？那个时候，用中心化的机构和方式，会是一个成本巨高，在商业上不可行的一个模式。

第二个问题还是以车路协同为例。我们怎么去保证现场两个设备可以在零点几秒之内完成数据交换，又如何知道设备有没有被黑客挟持呢？我们不时会在新闻中看到，国外发生了黑客挟持物联网设备的事件。

所以，我们首先要解决数据可信的问题。目前为止区块链是一个最佳的解决方案，最便宜也最适用。因为这么多设备无法都找中心化机构来问这是真的假的，或者中心化机构需要组织多大规模的一个团队来辨别真假？所以到那个时候，我们必须让所有的数据都发生在链上。

我们都知道区块链的非对称加密，目前为止还没有人破译过区块链上的私钥，所以黑客要去攻击链上的数据是一件很麻烦的事情。链上的数据又是一个公共账本，个人去修改账本是无效的。要串通这么多节点，成本高到惊人，这就在技术上保证了从"不作恶"到"不能作恶"。

目前区块链是没法作恶的网络，所以只要数据发生在链上，我们就可以相信这个数据，它几乎不可能是被改过的，也几乎不可能被删除。

数据怎么样发生在链上呢？这涉及物联网是基于互联网还是区块链。基于互联网上的数据有可能被黑客挟持，也有可能被设备的主人修改。但如果基于区块链，物联网的传感器被写在区块链上，那么数据就是通过链发出来的，就不可能被挟持或被修改。

信风智库：在数据资产化与数据要素市场化流通等方面，区块链与隐私计算等相关技术是如何发挥作用的？

肖风：数据资产化、数据流通和区块链是一个天然的绝配，这里面有三个问题。首先数据必须是真实的，否则大数据怎么可信，人工智能算法怎么训练？前面我们谈到了，让数据发生在链上就可以解决这个问题。

第二个问题就是数据权属的确认。当然目前有中心化的方法来确定数据权属，这个是没问题的。可是面向未来想一想，1000亿台设备都在发送和接收数据的时候，我们怎么用一套中心化的方法来确定这个数据归谁？来确定它是由谁产生的，应该怎么分钱，或者贡献数据后获得怎样的回报？中心化的方法到那个时候就变成一个极其昂贵的方案了。

所以，我们用区块链的方法来确权，加上运用像 NFT 这样的价值工具来为我们进行独有性的标注。区块链显然是一个很好的数据确权工具。

数据确权并不是最终目的。因为数据只有在交换当中才会有价值。你有数据我有数据，但我们没法交换，这个数据就沉睡在那里。数据在交换之前，就一定要解决第三个问题，隐私保护。

这就是隐私计算的用武之地。在加密状态下，我们再来共同计算某个想要得到的结果，让数据在交换过程中"可用不可见"。交换的同时还有一个清算的问题。区块链本质上就是一个新型的清算网络，除了数字货币交换要清算，数据的交换也需要清算。所以在数据交换方面，区块链还可以帮助我们做清算。

所以，在可信、确权与流通三个方面，区块链和大数据是绝配。

用数字货币服务去中心化世界

信风智库：之前您也谈到，中国推出了央行发行的主权数字货币。它将带来怎样的实际意义？将会产生怎样的影响？

肖风：当万物互联之后，面对这么海量的数据通信以及这么巨大的资源管理，我们可能在很多方面都需要用一个去中心化的技术架构才能支撑这么复杂的网络，这样才具备成本效益。那么，在这个基础上，经济也好，商业也好，也都需要去中心化。因为底座是去中心化的，就不能用中心化的方式来解决经济与商业上的问题，这样同样不具备成本效益。

在这个时候，不管是央行发行的主权数字货币，还是区块链上的加密数字货币，都是为了应对这种去中心化的世界以及分布式的技术、经济与商业。

如果不是为了应对这种状况，其实就没有必要设计数字货币，这是我的看法。因为我们如果完全不用考虑如何管理万物互联，不考虑如何治理去中心化的世界，我们目前的电子货币就已经足够了，尤其是在中国，我们已经有了一个很强大的电子支付网络。

所以我认为数字货币不是重复现有的电子货币，而是万物互联之后，我们要去治理中心化的世界，就需要基于区块链来重构一个与去中心化世界相匹配的新一代数字金融系统。

数字货币就是为此而生的，否则就没有意义。不管是主权数字货币，还是加密数字货币，最终的价值都在于治理这个去中心化的世界，而不是做一个新的电子支付系统。

这个去中心化的世界，根源就是万物互联，1000亿台设备需要端到端地、去中心化地做很多事情，做很多决策，做很多预测。比如就一辆汽车而言，要做很多预测，同时要做很多决策，这里面就涉及复杂且庞大的数据收集、交换与运算。

数字货币就是为此服务的，在去中心化的世界里，服务于去中心化的生产生活。它使用到的就是 Coin、Token 与 NFT，这个 Coin 可以是主权数字货币，也会用到加密数字货币。

信风智库：请您为我们推演一下，数字货币将如何逐步渗透到我们平常的生产生活中？

肖风：这取决于经济的数字化，生活的数字化，城市的数字化，企业的数字化。数字化进程因为新冠肺炎疫情开始突然加速。加上 5G 与物联网技术的发展，会进一步推动整个社会与经济的数字化进程。

因此，我们可以预计未来10年会出现一个去中心化的世界。当这种变化达到某种程度的时候，就会促使我们建立并制定与之匹配的技术、机制、模式与法律。去中心化世界生长得越快，这个过程也会随之加快。

我们现在已经可以看到很多创新的东西，其实也是为了服务那个生长得越来越快、变化得越来越大的去中心化世界。

其实，区块链技术本身也在发展。经过了 10 多年时间，区块链技术的基础架构到今天才能算是搭建完成。比如最近一两年产生的 Filecoin、Dfinity、Cosmos 与 Polkadot 这些新技术和新东西。

去中心化的世界，巨量的数据怎么存储？所以一定要有去中心化的存储，Filecoin 干的就是这个事情。这么多数据收集到了，怎么计算？全部发送到云计算的机房？那不现实。所以 Dfinity 这样的技术就出现了，它叫互联网计算机，没有中心化的机房，基于区块链在终端节点上完成计算，这就是去中心化的云计算。

去中心化的世界，有各种各样的区块链，各种区块链是彼此孤立、无法通信的，这就需要解决跨链和多链的问题。Cosmos 主要解决跨链通信的问题，让区块链之间可以相互通信与交易；Polkadot 主要解决多链融合的问题，将多个区块链融合到一个统一的区块链网络中。

加上大家熟知的加密数字货币提供的金融基础设施，到目前为止，基于区块链的整体技术框架才算搭建完成。

在这个基础上，我们还需要 3~5 年甚至 10 年时间去优化整个系统。搭建是一回事，优化又是另外一回事。搭建可以凭借创造力，优化就需要有具体用途。就像人工智能算法一样，算法可以由某个科学家写篇论文写出来，可是这个算法在某个场景里有没有效果，只有在具体场景里才能不断优化。

但这并不是说具体的创新应用非要等到 10 年之后才冒出来。它们可以不断地、逐渐地涌现。因为基于区块链的创业，门槛已经非常低了。

11个人写500行代码就干完了,创业的门槛低到这个程度,这意味着什么?意味着原来这个世界可能有100万个创业者,现在可以变成1000万个创业者。原来我找不着钱,没有资源,就无法创业。虽然我有个很好的想法,但资金门槛太高,技术门槛太高。而现在区块链把所有东西都下沉,成为链上的基础设施,基于上面的服务就只需要500行代码,技术门槛也降低了,资金门槛也降低了。这意味着创业者的人群就变大了。所以我们想想,未来10年将是怎样一个天翻地覆的10年。

罗熙文：
数字空间如何与现实世界共舞

罗熙文（Sylvain Laurent） 2008年加入达索系统，他始终致力于推动达索系统实现"走向市场"的战略转型。2011年至2019年，他担任达索系统全球直销业务执行副总裁兼亚太区总裁，负责全球大客户的产品全生命周期管理（PLM）、销售与解决方案，并管理亚洲市场。2020年，罗熙文被任命为达索系统基础设施和城市董事会主席以及达索系统执行委员会成员。

达索系统成立于1981年，是为全球客户提供3D体验解决方案的行业领导者。利用达索系统的3D体验平台和行业解决方案，科研机构、政府部门与企业可以对真实世界进行数字孪生，从而突破创新、学习和生产的界限。达索系统为140多个国家超过29万个不同行业、不同规模的客户带来价值，被评为"10家不为人熟知但却正在改变世界的欧洲公司"之一。

在中国，达索系统的3D体验解决方案已经在包括航空航天、汽车交通、船舶、工业设备、高科技、生命科学、能源与材料、城市建筑与地域开发等11个行业有成熟且深入的应用。

扫码观看
访谈精选视频

数字孪生的本质是什么？我们该如何正确理解数字孪生？数字孪生能否大幅降低现实世界的试错成本？我们发展数字孪生的目的究竟是什么？数字孪生将在"碳中和"中发挥怎样的作用？未来，数字孪生还能为我们带来哪些进步成果？

数字孪生降低试错成本

信风智库：达索系统在数字孪生领域耕耘已久，请您告诉我们数字孪生的本质是什么？我们该如何正确理解数字孪生？

罗熙文：正如您所说，其实数字孪生是一个已经存在很长一段时间的概念了。对于我们达索系统来说，数字孪生的应用最开始是在航空航天领域，我们给飞机做数字化样机。

从 1990 年开始，我们以数字化样机为载体，花了很长时间探索如何利用一个虚拟世界来改进航空航天这个行业的发展。这就是达索系统研究数字孪生的起点。

最近 5 年来，我们逐渐从数字孪生这个概念开始发展到了虚拟孪生。

虚拟孪生和数字孪生的区别，就在于虚拟孪生从原来单一的一个数字样机，进化到把一些相关的信息与功能集成到这个数字样机上面，让它变得更加丰富。

其实，虚拟孪生就是对现实世界的仿真模拟，这比 3D 建模的概念要广得多。我们已经把虚拟孪生的概念运用于各行各业。

达索系统 6 年前的"虚拟新加坡"项目，催化了我们虚拟孪生技术的广泛应用。其实现在各行各业都需要一个特别逼真的虚拟世界来更加真实地反映现实世界。对于我们来说，虚拟孪生就是唯一有效的反映方式。

信风智库：一个数字化的孪生世界或物体，对实体世界或实物究竟会起到什么作用，有怎样的益处？

罗熙文：这个答案其实是显而易见的。因为现在全球各个系统都变得更加复杂，上到各行各业，下到各个产品甚至人体，都是非常复杂的系统。

比如我们的"虚拟新加坡"项目。其实新加坡的城市是由无到有建立起来的，在过去的 50 年中，新加坡发展取得巨大成功，但是在下一个 50 年，新加坡应该采取怎样的发展方式？现在新加坡的领导人就需要利用虚拟孪生来对城市的复杂性进行管理，这样有助于他们做出更好的决策，比如该如何建设更加完善的基础设施、如何预测与应对公众安全问题等。

另外一个例子就是汽车行业。在过去，如果汽车制造商想开发一个新车型，就需要很多的测试车，测试加速、风阻、安全与碰撞等。经过这些极端测试后，大多数测试车都成了废品，必须被销毁。不仅成本高，也会浪费很多时间。

但是，我们现在 90% 的仿真和模拟测试都是可以利用虚拟孪生来完成的，这个过程也会更加安全。同时，还能针对车辆的不同性能，选择不同参数进行调试。所以其实我们可以利用虚拟孪生来发展新一代的车型，包括未来的无人驾驶汽车和网联汽车。

　　从这里可以看出，虚拟孪生不是简单的3D建模，它把物体的功能与性能连接在一起并反映出来。这个物体可以是一辆汽车，可以是人体，可以是城市，可以是飞机甚至是工厂。更重要的是，虚拟孪生不仅考虑物体本身，还会考虑它周围的环境因素，分析物体和周围环境因素是如何互相影响的。

　　信风智库： 创新最大的成本是试错。我们是否可以这样理解，即数字孪生可以大幅降低我们试错的成本？

　　罗熙文： 我觉得您的理解是非常准确的。人生没有彩排，但虚拟孪生就给了我们"彩排"的机会。一方面数字孪生或者虚拟孪生可以帮助我们缩短产品的整个研发周期；另一方面它可以为我们提供各种各样的方案。如果我们需要在多种方案或者多种场景之间做比较，它可以帮助我们减少成本，也可以降低风险。对于管理层来说，它还能加快做决策的效率。这些都是利用现实的实物与实体很难实现的。

　　除了可以缩短时间以及降低成本之外，就像刚刚强调过的，我们还可以把虚拟孪生放到各种现实使用场景中进行观察。比如汽车，你可以把它放在道路的场景中、城市的场景中，然后去观察这个车的行为，观察它在不同的情况下，应该满足怎样的条件才能达到舒适安全的行驶状态。这对于数字孪生来说也是非常便利的。

　　举个例子，我们可以通过数字孪生进行模拟测试：同样的汽车在北京、贵州、厦门的不同状态。比如厦门气候更加温暖一些，又是沿海城市，汽车的行驶表现肯定跟在其他城市的表现是不一样的。还有就是，不同地区的用户对汽车的期待也是不一样的，所以我们可以利用虚拟孪生在不同的场景下进行模拟。

连接实体经济与数字经济的桥梁

信风智库： 由于简单直观的体验方式而更易被人接受的数字孪生，可以助推各种数字化应用加快落地吗？

罗熙文： 我认为是这样的，因为现在的数字经济需要一些可信赖的仿真模拟技术作为支撑。无论是提供各种各样的产品还是解决方案，我们都需要数字孪生与虚拟孪生促进数字化应用与实际场景需求的融合。

在达索系统，我们管这种新的经济形势叫体验经济。因为我们认为未来需要给各种各样的用户提供数字化的体验，满足他们对数字化体验的期待。

所以我认为，我们必须要利用数字孪生与虚拟孪生技术来助推数字经济的发展。

信风智库： 从根本上讲，我们发展数字孪生的目的究竟是什么？

罗熙文： 无论是虚拟经济还是虚拟孪生，一个很重要的目的就是管理复杂性，包括局部的复杂性和全局的复杂性。这种复杂性包括各种各样的学科，也需要将各种各样的功能相结合。虚拟世界就可以提供一种简单的方式去管理这样的复杂性。

信风智库：数字经济与实体经济的融合是大势所趋。您认为数字孪生是连接数字经济与实体经济的一座桥梁吗？

罗熙文：我认为您的理解是完全正确的。我们可以利用虚拟孪生，以更简单、更便利的方式去连接现实和数字这两种世界。当然它还是具有一定的挑战性，所以我们要给予虚拟孪生的数字模型更多的信任。

因为对于虚拟孪生的数字模型来说，每个人可能有不同的期待、不同的需求，所以它也要集成多种不同的信息。其实对我而言，虚拟孪生更像是一种新的语言。利用虚拟孪生，我们可以管理复杂的数据、信息、元素和模型，它可以在不同层级与各个方面用非常简单的方式展现给我们看，从而减少我们犯错的概率。

数字孪生背后是数据融通

信风智库：以您和达索系统的经验来看，要打造成功的数字孪生应用，有哪些关键点？

罗熙文：我们首先应该讲讲方法论。我们需要用怎样的方法论，来描述虚拟孪生的应用场景。在达索系统有这样一套方法论，叫作以价值为导向。达索系统拥有非常多的行业顾问，他们具有各行各业的丰富经验。他们通过以下四个步骤来贯彻我们的价值导向方法论，为客户带来服务和支持。

其实这四个步骤也是非常易于理解的。第一步就是价值评估，去评估现在所处的状态，了解清楚现在这件事情处于怎样的态势情况。第二步就是价值定义，定义到底要提供怎样的价值，也就是场景需求应该是怎样的。第三步就是价值承诺，我们向客户承诺我们将会交付怎样的价值。自然而然地到了第四步价值交付，将我们承诺的价值予以交付。

如果说不使用这一套方法论，我们可能也有办法实现一个数字孪生或者虚拟孪生应用，但是这样的虚拟孪生就缺少一个更加准确的定义，定义它应该达成什么样的目标。我们可以利用虚拟孪生来帮助人们去了解，我们能交付怎样的价值，我们能达到怎样的目标。

信风智库： 达索系统早在 2012 年就提出了 3D 体验战略，并随后推出了 3D 体验平台，请您为我们介绍一下它是如何为全球用户服务与赋能的？

罗熙文： 刚才我们说到虚拟孪生可以用来管理复杂且多样的信息和数据，同时也可以用于管理人员、部门以及生态系统。我们推出的 3D 体验平台其实是一个数字化的平台，能够帮助我们连接数字世界和现实世界，同时可以打破数据孤岛，连通所有的数据。

我们可以利用一个新概念——3D 体验平台，去实现把不同地区、不同国家与不同行业的人聚合在一个同样的虚拟空间中，交流共同感兴趣的话题，协同完成一个任务目标。

比如，我们要建立一个机械零件的 3D 模型，以前各个部门之间通过电子邮件或者传送文件包的方式沟通，过程中容易出现误解与偏差，效率非常低。但如果将跨部门、跨地域与跨供应链的相关人员都聚集在 3D 体验平台上，大家共同面对同一个机械零件的 3D 模型，就可以直观高效地沟通与协同工作。

更有价值的是在这个过程中，产生了很多跨学科、跨产业的虚拟孪生模型，比如城市建筑、生物医疗和汽车机械等，它们的核心技术是共通的。这些模型就又可以与平台上其他有相同需求的机构分享，提高大家效率的同时，也能让模型本身的价值更大。

所以，3D 体验平台可以用于实现不同人群之间的交流，它也可以管理非常海量的数据，因为我们所谓的孪生，不仅仅在于一段文字或一句话，它其实承载着各种不同的信息和数据。另外这个平台也可以解决安全性的问题，包括数据的安全性，以及到底是在哪个国家或者在哪个场景下，哪些人有权限来使用这些数据。

为什么要给这个平台取名叫 3D 体验平台？因为我们都知道，现实世界的各种物体以及我们人体都是三维的，所以我们的想法就是用虚拟的 3D 模型来描述、映射与体验现实世界。这就是刚才讲到的，我们现在进入了一个体验经济的时代，可以通过虚拟孪生来尝试各种可能性。

信风智库：刚才您提到了数据的聚合与管理，我们是否容易被表面炫酷的 3D 影像所吸引，却忽略了真正重要的是数据背后的融通与应用？

罗熙文：我们先回到您刚才说的虚拟孪生。与 3D 体验平台相连的虚拟孪生，其实可以把它比作一座冰山。我们看到的只是冰山一角，真正大量的有价值的数据隐藏在海面下，是看不见的。

刚才说到了，我们可以利用这样一个虚拟的 3D 体验平台，用简单的方式去管理复杂性的问题。比如，在制造领域，我们可以利用虚拟孪生来集成整个价值链上所有的相关信息和数据，便于工程师们更加高效便捷地协调工作。对于各行各业也都是如此。

信风智库：请您为我们分享一下，数字孪生为各行各业带来了怎样的改变？

罗熙文：我们可以看到，无论是在过去还是未来，数字孪生或虚拟孪生在各行各业都有着非常重要的应用。比如航空航天这个行业，如果没有虚拟孪生技术的话，我们很难真正高效地造一架飞机。因为飞机设计制造过程需要各种模拟测试，包括新材料以及新技术的应用。所以如果没有虚拟孪生技术，设计建造一架飞机就需要很长的时间。我们现在能做到把时间从

20 年缩短到 4 年。

再说汽车行业，在新型交通出行方式上，我们对汽车的理解不仅仅是关于它的车身、底盘或者发动机应该怎么设计，更关注的是车内和车外的互相连接。我们应该把汽车放在具体的应用环境中，无论是一个城市还是一条智能道路，去观察并进行一些测试。所以如果没有虚拟孪生，我们开发下一代汽车也不是不能做到，但是一定会更麻烦，会花更长的时间，利用虚拟孪生可以大大节省我们的时间和精力。

刚才我们提到的都是制造业的一些例子。其实在基础设施、智慧城市以及生命科学的方方面面都有这样的例子。我还想提一个点，就是如今全球都在强调的碳中和目标。

其实我们现在设计或构建一个产品的时候，都会考虑它对碳排放的影响。我们利用虚拟孪生就可以进行准确的预测，在生产制造的过程中，甚至在物流的过程中，会有怎样的碳排放影响。

我们谈到的这一切，都是为了给各行各业来带来价值，帮助它们实现创新。比如建筑行业，目前的建筑行业正在掀起新一轮的变革。未来对于新材料的应用，新的建筑方式，包括新的工厂建造方式，都会有一次革新。未来 5 年，我们可以看到大大小小的建筑公司都会向数字化转型。

还有生命科学这个行业，新冠肺炎疫情在全球肆虐，对于所有人来说都是一个坏消息。但是我们可以看到生命科学行业利用数字化技术来加快疫苗的研发，此前疫苗的研发周期是 10 年左右，但现在不到两年就已经研发出了疫苗。

这背后其实也是通过虚拟孪生技术，仿真模拟出了各种不同的大分子结构，通过算法模型从上百万种甚至更多的组合可能性当中，找出三四种药理最有效且毒性最低的大分子结构组合。

数字孪生使城市更美好

信风智库:您刚才谈到了碳中和。这是目前中国乃至全球都非常关注的话题,也是人类未来将实现的目标。数字孪生能在这个领域发挥怎样的作用呢?

罗熙文:数字孪生或虚拟孪生在碳中和方面,主要是起到统计、监测与预判的作用,相关的应用案例就太多了。比如,对于一个城市来说,决策者想知道按照目前的生产生活水平,整个碳排放总量有多少?如何才能做到精细化的统计呢?

我们知道二氧化碳的排放来源既有固定的工厂烟囱与建筑,也有动态的飞机与车辆等交通工具。所以我们就需要利用虚拟孪生建立一个模型,来管理这样的复杂性。比如,统计这个城市里有多少个烧煤的发电厂,有多少个生产的工厂,有多少人口与汽车保有量等,然后细化到每个区域单元格,每个小区或园区,甚至每个建筑。这样我们就可以知道,这个城市里,哪些区域是"重灾区",其碳排放的密度如何,还可以用不同颜色标注出来。

接下来就是预测如何解决。如果我们发现这个城市碳排放的大头是煤电厂,我们应该如何治理呢?是直接关停还是逐步关停,还是能源改造,这也是一个动态的决策。所以我们需要基于虚拟孪生建立一个预测模型，用数据来推演：如果决策者要求一年内碳排放降低 50%，是不是就意味着大部分的煤电厂都得关门，这会导致多少人失业，会对产业链上下游产生什么作用，又会给经济带来多少影响？基于这些相关因素的综合分析，我们再来推演选择，是分三年五步逐步关停，还是五年三步，或者用天然气、太阳能以及地热来替换，从而确定哪种方式是最优决策。

信风智库：从数字孪生的角度来看，您认为全球范围内对智慧城市的理解与实践还存在哪些误区？

罗熙文：其实在过去 10 年，我们发现全球各地都在讨论智慧城市这个概念，但是很少能真正地把智慧城市付诸实践。为什么会这样呢？原因就在于缺少一个城市的数字孪生模型。

现在大部分的政府管理部门都还在使用成堆的纸质文件和材料，虽然各个部门都有创新的想法与应用，但却很少能够以一种有效的方式把它们聚集在一起。

比如"虚拟新加坡"这个智慧城市项目，就是利用虚拟孪生实现了数据互联互通。这不仅仅是一个看起来很炫酷的 3D 模型，它是一个真正的孪生，通过数据帮助决策者更快更准确地做出决策。比如在能源利用、城市安全与基础设施等方面，我们可以模拟预测车辆在这条道路上会不会拥堵，居民在这里过马路是否安全且不会影响交通效率，某块区域应该配置多少电能供给来支撑未来发展等。

利用数字孪生模拟城市建设与运行过程，综合各种各样的实际情况与可能性，依据数据分析做出最优的决策，这是新加坡建设智慧城市的一个重要成就。我们也计划在中国的一些城市实行这样的项目。

其实，收集真实世界的数据已经不再是一个特别难的问题了，我们可以利用物联网进行数据收集，但是很难实现的一点就是预测未来可能发生的变化会对市民造成怎样的影响，这是虚拟孪生真正的核心价值所在。虚拟孪生就有这样的能力，可以帮助决策者观察分析数据，然后预测这个决策未来可能会产生怎样的影响。

数字孪生的中国发展与未来趋势

信风智库：在您看来，中国发展数字孪生应用有着哪些优势？达索系统是如何在中国市场布局的？

罗熙文：我认为中国各行各业的发展速度非常快。更进一步地说，我们需要加快数字经济与实体经济的融合，加快数字化技术的应用，才能让各行各业都以同样的步伐速度向前迈进。

在中国市场，达索系统就可以提供这样的支持，利用虚拟孪生与数字孪生为各行各业赋能。我们了解了中国的"十四五"规划，达索系统的发展目标其实和"十四五"规划是高度吻合的。比如"十四五"规划上提出了碳中和以及培养下一代的高科技人才这样的目标，这些目标和我们的目标都高度一致。

实际上，达索系统的身影早已出现在中国很多家喻户晓的项目背后。比如鸟巢国家体育场、CCTV新大楼、大兴机场、国产大飞机C919以及全球首艘无纸化建造的船舶"海巡160"。

近年来，我们发现中国的原创型创新越来越多，而原创型创新往往是没有参照的，这就导致了极大的不确定性。如何减少不确定性，降低创新的试错成本，虚拟孪生大有用武之地。比如，我们助力中国的航空航天事业，模拟火箭和卫星的设计、建造、发射与运行过程；助力中国大型船舶与高铁的制造；与蔚来、小鹏和理想等中国新能源汽车公司合作。

为了进一步加大在中国市场的发展力度，近期我们在重庆设立了智能制造创新中心，也是希望利用我们的虚拟孪生技术来赋能重庆的制造业，助力重庆打造智造重镇。

另外，令我印象非常深刻的一件事情是，抗击疫情期间，在建设雷神山医院的过程中，我们与中南建筑设计院合作，捐赠了软件帮助他们结合现场实施特点，讨论制定常规暖通安装方案，进行流体力学仿真计算，并根据计算结果最终确认对污染物浓度控制最好、能及时将病人呼出污染物送至排风口的解决方案。同时，还依据此仿真结果，给现场医护人员提出一些站位与操作等方面的建议。这不仅能最大限度地减少医护人员交叉感染，还能避免对医院外部的社区群众和周边环境造成影响。

最后，我们认为很重要的一点就是培养下一代的高科技人才。因为教育是一个很漫长的过程，所以我们希望能够更多地赋能学校与中小企业，让下一代人才也能够更快地接触到这样的数字世界，让具有创新意识的中小企业也能把自己的生态系统带到这样一个数字世界中来。

信风智库：数字孪生应用越来越广泛，外科手术都可以用数字孪生来模拟了。请您为我们畅想一下，数字孪生未来还能为我们带来什么？

罗熙文：其实，3D数字世界的发展非常迅速，在中国的发展更快。我相信，或许我们还能助力中国航天事业一起去探索火星。

对于数字孪生与虚拟孪生的未来，我想谈三点。第一点就是您刚才提到的外科手术。事实上我认为在未来 10 年，我们可以更好地预测与控制人体。其实我们可以把人体比作飞机这样的复杂系统，其中有很多数据，我们可以利用虚拟孪生技术，在病患需要做手术之前，就预测出他的身体可能出现了怎样的问题，由此可以采取措施避免疾病恶化。

在这个方面，我们还做了一件非常轰动的事情，就是通过虚拟孪生制造了人造心脏。其实我们很早就在研究，如何利用所谓的克隆技术制造出人类的器官，从而延长人类的寿命或解决很多目前无法治愈的疾病。

我们最显著的一个成果，就是用真正的人体干细胞培养出一个人造心脏，然后进行心脏移植手术，延长了患者的存活时间。这是人类科学的一个进步。

在这背后，其实是我们建立了一个心脏的虚拟孪生模型，包括心脏的整体结构、肌肉壁、血管与神经等所有组织，一比一真实还原了人类心脏器官。由此延展开来的应用就非常有价值。比如，在一次比较高危的心脏手术中，容易出现爆血管等种种风险。这种情况下，外科医生就可以通过虚拟孪生的心脏，综合血压、心率与血液黏稠度等体征数据，反复进行手术模拟，反复试错排除风险点，进而研究出一种最为稳妥安全的手术实操办法，大幅提高手术的成功率与安全性。

第二点就是下一代的基础设施建设和城市管理，比如上海和重庆都是很大的城市，如何帮助政府对这样的大城市做出更好的决策，也要依赖虚拟孪生技术。

第三点就是现在非常热门的碳中和目标，我认为不管是现在还是未来，虚拟孪生技术都可以用于保护地球环境。

李远清：
脑机智能如何孕育"最强大脑"

李远清 IEEE Fellow、国家杰出青年科学基金获得者、教育部长江学者特聘教授，广东省高等学校珠江学者特聘教授，琶洲实验室常务副主任，华南理工大学自动化科学与工程学院教授、博士生导师，华南理工大学脑机接口与脑信息处理研究中心主任，享受国务院特殊津贴专家，"百千万人才工程"国家级人选。

1997年毕业于华南理工大学自动控制工程系，获工学博士学位；1994年毕业于华南师范大学数学系，获理学硕士学位；1988年毕业于武汉大学数学系，获理学学士学位。曾在香港中文大学、香港城市大学、日本脑科学研究所、新加坡信息通信研究所神经信息处理实验室、英国华威(Warwick)大学访问研究。致力于信号稀疏表示、盲信号分离、半监督机器学习、脑机接口、脑电与功能核磁共振成像信号分析等方面的研究。学术兼职包括 *IEEE Transactions on Fuzzy Systems*, *IEEE Transactions on Human Machine Systems* 等4份SCI国际期刊的副主编、中国自动化学会常务理事、中国人工智能学会理事等。

近年来，主持项目包括国家重点研发计划项目、国家自然科学基金重点项目、广东省脑计划项目等。相关成果获国家自然科学二等奖1项、教育部自然科学一等奖1项、广东省自然科学一等奖2项等。

扫码观看
访谈精选视频

如何理解脑科学？智能时代赋予了脑科学哪些新意义？全球各国的脑科学研究发展处于怎样的阶段与水平？中国提出"一体两翼"的脑计划具体是什么，"一体两翼"之间又如何关联与协同？脑机接口下一步的产业化应用，将主要聚焦在哪些方面，还需要解决哪些问题？

从"理解脑"到"保护脑"与"创造脑"

信风智库：在智能时代，我们应该如何理解脑科学，智能时代赋予了脑科学哪些新的意义？

李远清：人工智能发展的一个很重要的目标是实现通用人工智能，通用人工智能跟人类大脑的结构原理是非常相似的。大脑的本质就是处理内外部的信息，比如内在的经验与知识，以及外部的感知，然后做出相应的反应和决策。而人工智能也应该像大脑一样，能够处理自然界的非标准化信息，从而进行自动化反应和智能化决策，而不仅仅体现于数据统计、打游戏、下棋或者超强的记忆力。从这个角度讲，人工智能和脑科学的发展是互为补充、相互促进的。

一方面，脑科学的发展能给人工智能提供新的方法。我们越理解大脑运转的逻辑，就越能将各种零散的人工智能技术整合起来，从而促进人工智能的模型、算法和产品进一步提升，打造更智能的信息处理系统，甚至达到人脑的能力。这是脑科学对人工智能发展的推动意义。

另一方面，人工智能的新技术和新方法也可以反哺脑科学的研究。人类社会所有的活动都是大脑运作的结果，所以脑科学一直致力于破解大脑工作的原理，比如大脑是如何做决策的？什么是意识？为什么会有创新？为什么有不同的性格？例如，人工智能不仅能通过脑机接口技术采集大脑数据，还能通过算法模型对大脑数据进行分析和解码，从而帮助我们进一步理解大脑的运作逻辑和决策机制，助推脑科学研究。

实际上，现在我们对大脑的研究已经有了一些突破，特别是在一些脑疾病诊疗方面，例如通过人工智能的方法对脑疾病实现早期诊断或干预治疗。

信风智库：从全球范围来看，各国脑科学研究发展情况如何？

李远清：毫无疑问，未来脑科学将为社会发展带来深远的影响，也已经成为全球各国科技竞争的战略高地。

美国白宫早在2013年就提出了"脑计划"，旨在探索人类大脑工作机制、绘制脑活动全图、推动神经科学研究以及针对目前无法治愈的脑疾病开发新疗法。同年，欧盟聚集了不同领域的400多名研究人员，设立了专项计划——"人脑计划（HBP）"，希望通过计算机技术模拟大脑，建立一套全新的和革命性的生成、分析、整合与模拟数据的信息通信技术平台，促进研究成果的应用落地。日本的"脑计划"则侧重于医学领域，主要是以猕猴大脑为模型，进行人类脑疾病的研究，比如老年性痴呆和精神分裂症等。

而中国的"脑科学与类脑科学研究"已经被列入国家重大科技创新和工程项目，并提出了"一体两翼"的布局规划。中国科学院在 2021 年初成立了包含 20 个院所以及 80 个精英实验室的脑科学和智能技术卓越创新中心，致力于在脑疾病早期诊断与干预、脑科学以及类脑智能器件三个前沿领域的深入研究。可以看到，相比于欧美国家，中国的脑科学研究虽然起步相对较晚，但基于扎实的科研基础，成长速度较快。

我特别强调一下脑机接口技术，这种技术的核心是采集大脑信号，并进行解码和处理，然后生成一系列指令，从而实现大脑与外部设备的直接交互。目前，我们正尝试将脑机接口技术与各种传统场景相结合，希望能解决医疗康养、文化和安全等多个领域的难题。

信风智库：中国提出"一体两翼"的脑计划具体是什么，"一体两翼"之间又如何关联与协同？

李远清：首先说"一体"，就是要持续地认知和理解大脑，这是基础。大脑内部就像一个浩瀚的宇宙，拥有太多的未解之谜，如果我们能更好地理解大脑的运作逻辑，就能够更好地理解整个世界。比如"全脑介观神经联接图谱"大科学计划就希望通过将大脑神经元细胞的连接图绘制出来，从而帮助我们更好地理解大脑，就好比我们要解析计算机里的芯片，一定要把里面的线路连接搞清楚，才知道它是怎么处理信息的。所以这项计划就是"理解脑"的一个典型研究方向。

再说"两翼"。在理解了大脑本质的基础之上，我们就可以更好地保护大脑，以及模拟大脑甚至创造大脑。

"两翼"当中的其中一翼，就是保护大脑。它需要研究脑疾病的诊断与治疗，并形成各种新型的医疗产业。大脑对于人类的重要性不言而喻，如何保护大脑，防止大脑的衰退以及脑疾病的产生等意义重大。"两翼"当中的另一翼则是模拟大脑或者创造大脑。它主攻类脑人工智能、类脑计算、脑机接口等与人工智能相关的新技术研究，希望能够创造出具有人类智慧的机器，这是未来人工智能发展的重要目标。

"一体两翼"的核心就是"理解脑""保护脑"和"创造脑"，这也对应了脑科学研究的三大方向。"一体"是基础与支撑，而"两翼"是发展与延伸。我们只有深度认识与理解了大脑后，才能去提升大脑的功能，保护大脑以及创造大脑，所以它们之间是相辅相成的。

人工智能如何学习人脑

信风智库： 脑科学是一项综合性非常强的研究，它需要哪些学科交叉融合来提供支撑？

李远清： 脑科学是一个涉及广泛、极为复杂的研究方向。如果狭义的理解，脑科学就是神经科学，它研究神经系统内分子水平、细胞水平、细胞间的变化过程，以及这些过程在中枢功能控制系统内的整合作用。而美国神经科学对脑科学的广义定义则是研究脑的结构和功能的科学，还包括认知神经科学等。总体而言，脑科学是高度交叉融合的专业，需要数学、物理、化学、计算机、信息学、心理学、生物学和医学等领域的背景知识。在整个脑科学的研究中，这些学科之间是交叉、协同、融合与相互支撑的。举个例子，虽然医学主要侧重于"保护脑"方面的研究，但医学本身拥有海量与大脑相关的医学数据，这能为"创造脑"的研究提供基础数据支撑。

实际上，脑科学的研究和发展，不仅需要各个学科之间的交叉融合，更要以应用场景落地为导向，这样一方面可以更好推动脑科学的研究，另一方面也可以用脑科学的研究成果造福人类。比如人工智能的人脸识别技术就是源自加拿大心理学家赫布提出的"细胞集群假说"，它可以简单概括为，当处在大脑不同部位的一群神经细胞存储着一段记忆时，如果只给出部分记忆刺激，这个集群的所有神经细胞都会被激活，从而实现对比识别和再记忆。

怎样连接人脑和计算机

信风智库: 从脑电波到脑信号再到电信号,脑机接口的原理逻辑是什么?其中又有哪些技术关键点?

李远清: 脑机接口,顾名思义,就是"大脑+机器+接口",它通过接口让大脑与机器设备直接建立一个新的信息通道。脑机接口的工作原理可以分为四个步骤:信号采集、信号处理、控制设备和信息反馈。具体来说,就是脑机接口技术通过信号采集设备从大脑皮层采集脑电信号,经过处理,转换为可以被计算机识别的信号,然后提取当中的特征信号,再利用这些特征进行模式识别,最后转化为具体指令,实现对外部设备的控制。

所以脑机接口的本质是在人类大脑与现实世界之间建立一种全新的通信技术和控制技术。举个例子,人的行为都是由大脑发出指令,身体器官接收指令然后才做出反应,比如走路、跑步和打球等各种运动。如果人体的四肢出现了严重的损坏,就会丧失运动功能。在这种情况下,只要大脑还是正常运转的,我们就能为病人装上义肢,通过脑机接口技术实现大脑控制义肢,从而帮助病人恢复运动能力。

那么脑机接口技术的关键点在哪里呢?首先是大脑信息的采集。大脑的核心功能区在皮质层,皮质层的各个部分负责处理听觉、视觉及各种感觉信息,同时还掌管着语言、运动、思考、计划和性格等诸多方面。而脑机接口的信息采集主要是对颅骨下的内脑皮质层信号进行提取，所以接口方式与技术尤为关键，既要获得高质量的脑信号，又要保证大脑本身不被损坏。

再说反馈，大脑思考和决策很重要的一个环节就是获取反馈。我们通过感知能力去感受环境，并且将感知信息传递给大脑进行反馈，比如物体硬还是软，环境热还是冷，颜色红还是绿，大脑反馈信息后才能进行下一步的处理与决策。这个过程对于人体感知来说很简单，但是对于脑机接口而言，其实极为复杂。因为脑机接口要将外部机器设备获得的环境信息反馈给大脑，其路径和机制跟人体感知完全不同，它需要运用多模态感知的混合解析技术。

举个简单的例子，外部机器虽然可以感知到一个物体，但它却很难分辨这个物体的物理属性，比如"柔软"与"坚硬"。而且这种物理属性之间也是有关联的，当一个物体同时具有"坚硬"与"光滑"的属性时，它往往也会有"冰冷"的属性，这就是硬度与温度的通感，这种复杂的触觉属性对于技术感知而言极为困难。而多模态感知的混合解析技术就是通过构建触觉感知计算模型，感知物体的物理属性，并把相关的属性进行关联分析，实现通感并将实际感知传递给大脑。就目前而言，这是极具挑战性的技术。

信风智库：非侵入式与侵入式在技术原理上有怎样的异同？两者又各有哪些优劣势？

李远清：目前脑机接口技术主要有非侵入式和侵入式两种。非侵入式是外接接口的方式，就是指无须进入大脑，只需将穿戴设备附着在头皮上，实现对大脑信息的记录和解读。这种方式的好处是，它可以避免开颅手术造成大脑损伤的风险。缺点则是外接设备距离皮质层较远,中间隔着颅骨,所以脑电波信号采集受周围环境噪声影响较大,导致信号处理的质量和精度较低。

其实,脑电图就是典型的非侵入式系统,虽然它的易用性、简便性和性价比都很可观,但脑电图技术对噪声很敏感,精度控制上容易受到影响。

而侵入式则是通过手术等方式,将电极植入脑内,从而获得高质量的脑电波信号。这种植入方式对大脑的干预和调控更为直接,让大脑控制外部机器更加方便,但由于要做手术,这就存在相对较高的风险和成本;另外,异物侵入可能会引发大脑免疫和愈伤组织的反应,从而导致电极信号质量衰退甚至消失,伤口也难以愈合,甚至出现炎症反应。现在商业化的脑机接口几乎全都是非侵入式的,侵入式脑机接口则主要应用于特殊病人的治疗研究上。不过未来随着植入技术的进一步提升,侵入式的脑机接口应用前景也是非常广阔的。

信风智库:对非侵入式和侵入式两种模式而言,未来的主流探索方向将是怎样的?分别在哪些领域产生应用?

李远清:实际上,侵入式和非侵入式的脑机接口技术没有优劣之分,它们是两种完全不同的模式,都有其合适的应用领域。侵入式脑机接口需要做植入手术,它可能对大脑产生伤害,但是脑电波信号的采集质量以及信号处理的质量更佳,在神经调控方面更直接有效。所以侵入式的脑机接口主要应用于病人的诊疗,比如癫痫、阿尔茨海默病、帕金森病或者一些精神疾病等。

马斯克的 Neuralink 公司就聚焦于侵入式脑机接口的研究。虽然目前这项研究还没有正式应用于人的大脑,但 Neuralink

公司已经开始动物实验，并取得了很好的成果。2021年2月，Neuralink成功地让一只猴子通过植入大脑芯片实现了玩乒乓球的电子游戏。其方式就是将2000多个电极植入猴子大脑两侧的运动皮层中负责协调手部和手臂运动的区域，实现了对猴子大脑电信号的记录和解码。当猴子试图玩游戏时，大脑神经元产生的脑电波就被脑机接口设备读取，它会记录下哪些神经元被激活，并将相关数据发送到计算机的解码器。利用这些解码的数据，研究人员可以围绕猴子的神经活动模式和操纵游戏杆的手部动作之间的关系，建立一个数学模型，并对解码器进行校准，从而通过猴子的大脑活动来预测其手部动作。一开始，猴子接受的训练是用右手操纵游戏杆玩游戏，随着猴子的大脑意识与解码器同步，断开游戏杆后，猴子照样能够想象手部动作来操控游戏当中的移动指令。

侵入式脑机接口技术不仅在猴子的实验上取得了成功，在人身上的实验也取得了重大突破。例如 *Nature* 最近发布了一项侵入式脑机接口研究，可将人脑中的"笔迹"转为屏幕字句，速度和准确率都非常高。

而非侵入式脑机接口对大脑没有伤害，但是它对大脑的干预没有那么直接和精准，所以非侵入式技术更多地应用于教育、娱乐与智能家居等方面。

比如在教育上，我们可以开发很多脑机接口应用系统，包括制定个性化的训练方案、训练注意力以及缓解学习压力等；比如在娱乐上，非侵入式脑机接口技术可以与虚拟现实技术结合，我们就不再需要额外的操控设备，直接通过意念来控制游戏角色，在游戏世界中获得更加沉浸式的游戏体验；又比如在智能家居上，脑机接口技术与物联网结合，可以扮演"遥控器"的角色，帮助用户通过意念控制门窗、电视、空调以及灯光等，进一步提升居家的科技感与便捷性。

脑机接口的产业化风口在哪里

信风智库： 脑机接口下一步的产业化应用，将主要聚焦在哪些方面？还需要解决哪些问题？

李远清： 从已有的一些商业化落地应用以及未来产业化探索角度来看，脑机接口将会向三个方向发展：状态识别与监测、信息交流与控制、感知与运动功能康复与增强。

第一个方向是状态识别与监测。脑机接口正在切入教育、文娱和工作管理等领域。在教育领域，脑机接口设备可以对学生的注意力进行实时评测，帮助教师获得教学效果的实时反馈，并为改进教学内容提供参考。在工作管理上，比如航空飞行员、空中交通管制员以及长途货车司机等特殊岗位人员如果出现疲劳状态就会影响安全，而脑机接口可以应用于实时监测这些特殊岗位人员的大脑数据，为工作管理提供客观依据。

在这些商业落地场景中，非侵入式脑机接口方案将会是主流，而相关产品则是便携式的可穿戴设备。以往，脑电波信号的获取必须依赖复杂的仪器设备，要求被测试者佩戴笨重的、布满线缆的帽子。而未来随着新材料与信号处理算法等方面的突破，便携式的可穿戴脑机设备会越来越成熟。

第二个方向是信息交流与控制。在这个方面，我们看到越来越多的脑控外部设备逐渐问世，比如脑控机械臂、假肢和遥控器等，其核心就是通过脑机接口实现大脑直接控制机器设备。哈佛大学研究团队发明了一款基于脑机接口技术的智能肌电假肢——BrainRobotics。该假肢采用人工智能算法处理脑电波和肌电信号，通过表面肌电传感器检测残疾患者残余肢体的肌肉活动，从而训练患者通过主动收缩肌肉让假肢做出多种动作。基于这样的技术原理，患者可以实现像写毛笔字、弹钢琴和攀岩这类难度极高的行为动作。

在这一方向上全球最新的研究成果是 2020 年美国约翰霍普金斯大学宣布首次实现同时控制两条机械臂。借助脑机接口同时控制两条机械臂是一项比较困难的挑战，因为它不是简单的"左臂算法 + 右臂算法"，而是要通过目标任务进行总体的认知与解析，再分配给双臂共同完成。它需要将更多的统筹性、关联性与协调性融入算法模型当中，所以用脑机接口同时控制两条机械臂的难度，比控制单臂要难好几倍。

第三个方向是感知与运动功能康复与增强。在 2014 年巴西世界杯的开幕式上，一个全身瘫痪的年轻人在脑机接口和机械外骨骼的帮助下，成功地踢出了那届世界杯的第一球。

在这之前，这名年轻人的瘫痪程度很高，只有 4 节脊椎能够活动，而借助脑机接口技术将运动意识传递给机械外骨骼，通过外骨骼带动肢体运动，经过 10 个月左右的训练，成功让 11 节脊椎恢复了感知和运动控制功能，可以说从"完全瘫痪"修复成了"部分瘫痪"。

未来，"脑机接口 + 机械外骨骼"的组合康复方案以及相关产品将拥有巨大的应用价值和市场潜力，可以为全球上千万的患者带来希望。

这是脑机接口未来产业化的三个方向，但是同样也面临一些问题。一方面，如何解决用户接受度的问题。无论是侵入式还是非侵入式的脑机接口，对于普通用户来说，目前都不具备很高的接受度。因为这项技术跟大脑强相关，其安全性还没有得到大部分人的认可，或者说还没有形成一个官方或行业的标准，所以现在的应用大多都是针对极个别特殊患者。这是脑机接口技术必须提升的一个重点。

另一方面，就是产品化过程中性能与成本之间的平衡。既要保证其功能应用能够切实解决用户需求，毕竟对于新的技术产品而言，用户需要看到直观的效果才会买单；又要平衡企业自身技术设备研发与生产的成本，如果不能持续盈利，产业化也无法走得更远。

信风智库： 医疗是脑机接口的重要应用领域之一，请您为我们介绍一下，目前在医疗领域形成了哪些解决方案与应用成果？

李远清： 先举一个意念打字的案例。国外有研究团队开发出了一种侵入式脑机接口，成功让一位65岁的瘫痪患者实现了用意念书写字母，然后在计算机屏幕上实时显示这些字母的打字版本，而且准确率极高。这是如何做到的呢？工作原理就是通过在患者脑内植入两个微小的电极阵列，用来采集和接收大脑控制手写字母的信息，也就是大脑中想象手写字母的意念，然后构建一个算法模型，对相关的大脑信息进行解码与转化，形成字母的图像显示在屏幕上。

实际上，这项技术的底层算法逻辑就是以机器学习为基础。意念手写字母需要进行大量的训练，才能进一步提升精准度。特别是像"2"和"Z"这样形状相近的字符，手写结果难以区分，这就需要算法模型进行大量的前后文训练比对，比如"2"的前后都是数字，"Z"的前后都是字母，从而作出判断。

再举国内第一个侵入式脑机接口辅助治疗的案例。一位72岁高龄的患者，在几年前遭遇车祸，四肢完全瘫痪，只有大脑完整并未受损。浙江大学医学院附属第二医院通过侵入式脑机接口，成功帮助这位患者实现了利用脑电信号来精准控制外部机械臂与机械手，恢复了部分行动能力。

这个过程其实并不容易实现。虽然通过侵入式的脑机接口可以实时记录患者大脑发出的脑电信号，但是控制机械臂却很不容易，因为它需要脑电信号发出的指令与机械臂的动作充分磨合与协调。比如患者在想象控制机械臂移动时，如何控制摆幅与力度，精准地让机械臂到达自己指定的位置，都需要反复进行机器学习和人工训练。

浙江大学研究团队采用了自主研发的人工智能算法来解码患者发出的脑电信号，指挥机械臂进行更自然顺畅的运动，同时这位高龄患者也经过4个月的交互训练才实现了流畅的意念控制。现在，该患者已经可以用意念控制机械臂从桌上拿可乐，并插上吸管精准地送到嘴边了。

另外，我们华南脑控智能科技有限公司在医疗相关领域也有一些成果应用。比如意识检测，针对大脑有严重损伤的患者，进行视觉刺激或者听觉刺激，然后用脑机接口对患者的反应进行检测，评定患者是否具有人脸识别、数字认知、情绪表达等方面的功能，进而判断患者的意识障碍程度。还有一个应用就是运动功能辅助，对于那些缺乏行动能力的患者，我们开发了一套以脑机AI技术为核心的智慧病房系统，它能有效保证病人通过头戴式脑电交互设备向护士发出呼叫指令，还能实现对灯光、空调、护理床等设备的简单控制。

　　脑机接口技术在医疗领域内的应用价值极大，除了对身体运动功能障碍的患者具有康复治疗效果以外，在精神障碍患者的检测、辅助诊断以及康复预测等方面，都具有极大的潜力。这些方向上的技术研究其实已经取得了一些成果，相信不久的将来，能够更多地落地到实际应用当中。

脑机接口还有哪些想象空间

信风智库：现阶段脑机接口技术的发展还存在哪些问题？未来脑机智能又将如何突破？

李远清：脑机接口是一个新兴的多学科交叉领域，虽然在现阶段取得了一些比较重要的突破，比如辅助一些人类疾病的诊疗，但还有很多研究依然停留在实验室阶段，距离大规模的普及应用还存在相当遥远的距离。

脑机接口技术进一步的应用发展还需要突破以下三个方面的瓶颈。首先，最重要和最难的一点，就是脑科学的研究还没有实质性的突破，人类对大脑的理解还非常浅薄。即使大家已经认识到大脑能够发出指令控制行为，但不了解指令形成的过程，也就无法有效读取指令。举个例子，我们使用鼠标和键盘，需要手臂和十根手指相互配合，以合理的时间、位置和力度去操作，最后达到我们想要的结果。从实际的动作看，很多人可能觉得打字特别简单，就动动手指头。但如果究其运作逻辑，我们会发现，大脑里参与这一过程的神经元有很多，并且遍布不同的脑区，它们的信号非常复杂，采集完整信号极为困难，而解码更是难上加难。以目前的技术，我们最多只能在特定脑区采集部分神经元信号，然后结合最新的数据分析方法，希望能够以小见大，破解大脑的信号指令，所以精准度不能做到100%。

其次，脑机接口的硬件设备研发还存在一些挑战。由于要对大脑信号进行采集，所以脑机接口需要特定的硬件设备。一方面，需要脑机接口的硬件接入尽量不伤害大脑；另一方面，又需要硬件设备尽可能小巧便捷、功能稳定和性价比高。显然，目前的脑机接口设备材料只能满足其中部分条件，还有很大的提升和研发空间。

最后，脑机接口目前的应用范围还局限于医疗领域，对普通的生活领域涉猎较少。虽然进行了一些针对用户应用的研究，比如通过脑机接口操控鼠标以及玩游戏等，但还存在诸多限制条件，并没有完全进入日常生活。如何开发出一些能引起大众兴趣的应用，还需要研发机构和科技公司持续探索。

总的来说，脑机接口作为一个新兴领域，拥有无限的想象空间，实现人脑与计算机（人工智能）的连接只是第一步。

信风智库：随着脑机接口技术的不断发展与突破，意识存储、意识读取与人机共生等概念似乎变得可能。对于这些设想，我们应该如何理解与看待？

李远清：实际上，我们所说的"读心术"或者"心灵感应"，目前的脑机接口技术在一定程度上已经实现了。比如之前提到的案例——通过脑机接口实现"意念打字"，从本质上看，这就是一种"读心术",我们心中在想什么,脑机接口都能将它表达出来。

毫无疑问,对于技术发展本身而言,"读心术"一定是未来的研发方向,而这种技术的深化也会带来变革性的正向价值。比如我们能通过这项技术提前识别或预警谁患有精神分裂症、阅读障碍、孤独症或阿尔茨海默病,还能查看大脑的活动模式来确定一个人是否有自杀倾向,从而及时进行干预和挽救;又比如,通过脑机接口技术了解人类的恐惧、焦虑、吸引与愉悦等情绪感知系统,我们就能创造出更符合用户需求的产品与服务,全方位提升人们的幸福感。

同时,我们应当注意在正向价值的背后,通常也会伴有负面影响,一项新技术在创新过程中通常都会面临道德伦理问题。大数据带来经济价值的同时,也带来了数据隐私和安全的问题。而脑机接口技术如果进一步深化发展,无疑也会面临类似的问题,届时就需要政府部门、科研机构与产业界等来共同制定相关的法规和标准。当然,就目前的发展阶段而言,脑机接口还处于技术探索的初期,重点还是在技术创新、迭代与成熟上。

至于大家谈论的"意识输入和存储",只是一种畅想或一个概念,距离实现还非常遥远。不过,从宏观的角度来说,实现人类智慧与人工智能的融合是不少研究机构的远景目标。

部分国内外的脑机接口公司

公司名称	国家	技术路线	业务方向
浙大西投脑机智能科技	中国	侵入式/非侵入式	成为脑机关键技术与产品原型研发基地,构建"脑-机-脑"双向交互开放国家级平台,践行产研深度融合,坚持产研深度融合,坚持需求导向,助推脑机智能相关产业发展
华南脑控智能科技	中国	非侵入式	研发方向主要聚焦于教育和医疗领域
科斗脑机科技	中国	非侵入式	脑机接口相关软硬件开发,产品主要应用于医疗和教育领域
布润科技	中国	非侵入式	脑电采集、脑电分析、脑电模式识别、脑机接口控制与应用
博睿康科技	中国	非侵入式	致力于神经科学创新研究和临床神经疾病诊断、治疗与康复研究提供专业、完整的解决方案
脑陆科技	中国	非侵入式	专注于脑科学、脑健康筛查、脑电算法、脑电数据开放平台、类脑决策计算等脑科学前沿科技应用
NeuraLink	美国	侵入式	开发高带宽且安全可靠的脑机接口技术
Brainco	美国	非侵入式	致力于开发非侵入式脑机接口技术,研发基于脑电波的可穿戴设备,达到注意力训练、半瘫患者功能恢复等目标
Kernel	美国	侵入式	研究人类智能,致力于研发神经义肢技术
NeuroSky	美国	非侵入式	研发脑电波控制的可穿戴设备
Emotiv	美国	非侵入式	研发简单易用的移动可穿戴设备,可监测用户的认知和情绪状态
Mindmaze	瑞士	非侵入式	结合VR、脑成像、计算机图形学和神经科学的平台
InterXon	加拿大	非侵入式	开发一种脑波监测头环,可通过实时音频反馈帮助用户冥想和放松

(信凤智库根据网络资料整理)

谢更新：
月球上的第一片绿叶意味着什么

谢更新 重庆大学环境与生态学院教授、博士生导师，教育部深空探测联合研究中心副主任、重庆大学先进技术研究院院长、教育部科技委学部委员，探月工程嫦娥四号任务突出贡献者。教育部创新团队发展计划团队带头人，担任我国嫦娥四号生物科普试验载荷总设计师。国际宇航联合会教育专委会成员，中国宇航学会理事，中国高等教育学会军民融合教育研究分会理事。中国宇航学会深空探测技术专业委员会委员，曾担任国家载人航天与探月工程等国家相关专业组专家。

2013年获得探月工程三期关键技术攻关和方案研制优秀个人；2009年获得教育部技术发明奖一等奖（排名第三）；2003年获得教育部科技进步奖一等奖（排名第三）；2006年获得国家科技进步奖二等奖（排名第六）。

扫码观看
访谈精选视频

深空探测的主要目的是什么？它对我们的国民生产生活有哪些影响和意义？深空探测和数据收集与运用，有着怎样的关联，数据如何促进这方面的科研？大数据智能化等新一代信息技术，如何运用到深空探测领域，深空探测又如何拉动大数据智能化技术发展？

揭开月球背面的神秘面纱

信风智库： 嫦娥四号是人类历史上第一个在月球背面着陆与巡视勘察的探测器。迄今为止它已经取得了哪些重要的科研成果和成就？

谢更新： 嫦娥四号在月球背面着陆，取得了一系列成就。从 2019 年 1 月 3 日一直工作到现在，嫦娥四号获得了月球上的温度、太阳风等相关环境数据，还有月球表面的地质地貌和地质结构的数据。大量的第一手资料，为下一步构建月球科研站以及月球基地，进行有人的月球探索奠定了很好的基础，也为科学家研究月球的起源以及月球资源的开发与利用，起到了一个很好的科学指导作用。

信风智库：嫦娥四号在月球背面探索与以前在月球正面探索，这两者有怎样的区别？

谢更新：其实这不是人类首次对月球背面进行观测，因为很多飞行器也绕着月球转，也在进行一些观测。只是这是第一次航天器落在月球背面上进行原位就位的探测。

月球背面其实和正面差不多，只是以前人类是通过观测来探测月球，认为月球背面很神秘，所以当时月球背面也被翻译成"dark side"，因为从地球上是看不见它的。我们都知道月球在自转的同时也围绕地球公转，由于潮汐锁定现象，所以在地球上观测只能看到正面，不能看到背面。

随着技术的发展，我们在月球轨道设置了中继星——鹊桥，把月球背面的图像数据与信号等通过中继星传回地球，我们发现月球背面和正面没有根本性的区别。所以现在我们把月球背面叫作"far side"，意思是比正面离地球稍远的一面。

月球上的第一片绿叶

信风智库：在嫦娥四号上，您主导的生物科普试验载荷项目，生长出了月球上的第一片绿叶。从深空探测以及生物生态的角度来看，它具有哪些意义？

谢更新：在这个生物实验中，我们构建的密闭生态微环境，除了有大气压的空气和适宜的温度与湿度之外，其他条件都是力求保持月球表面的真实情况，比如月球六分之一的低重力场、月球上太阳自然光的强光照以及没有任何防辐射措施等。这样一种特殊的空间环境是我们在地球上无法完全模拟的。

那么，在月球非常特殊的环境下，生命能否适应环境？如何适应环境？能否生长以及怎么生长？都具有非常大的不确定性，所以非常值得我们去验证。这就是第一个意义，因此 Nature 评价我们的试验为 "pioneering experiment"（开创性的实验）。

就像 NASA 生物方面的首席科学家 Mike 教授说的那样，我们把种子带到月球上，如果 14 天不死亡，那么就意味着种子在月球上是可以生长的。在我们的实验中，种子不但没有死亡，还非常幸运地长出了一片绿色的叶子。因此生物在这种特殊环境下通过构建适宜生存的环境是能够适应且能够生长的。

我们做这个实验，用罐子密封了地球上的空气、土壤和水，但还有几个条件是没有刻意去营造的。比如，我们让月球上的太阳光直接照进罐子里进行光合作用，在罐体材料上没有考虑光照强度，没有考虑防辐射，保持了真实的光照条件。所以下一步我们将考虑更多地利用月球上的原始条件，比如利用月球土壤，甚至在月球上造水。目前在月球上原位造水的技术，大家都在跟进，也有了一定积累和可行性。这个实验的原创性受到了国际上充分的肯定，*Science* 评价我们的成果是"in a first for humankind"（人类的第一次）。

第二个意义就是展示性。我们在设计这个实验的时候，充分考虑了它的展示性，一定要把种子在月球上面的生长过程拍成照片并传回地面。除了科研价值之外，还可以让广大青少年了解我们在月球上是怎样去营造这样一个特殊环境以及在这样的特殊环境下生物是怎样生长的，从而激发青少年的科学探索热情。同时，这么多工程师和科学家努力做这个实验，花了这么大的代价，在月球上长出了一片叶子，可以唤醒广大民众的环境保护意识，让大家知道地球环境保护的重要性。我们团队已经在国内大中小学开展科普讲座 120 多场，并受邀到英国、西班牙、法国与瑞典开展交流，传播了科学知识并宣传了我国探月工程的成果，受到了大家好评。

第三个意义就是工程实现。我们用了非常小的代价来做这样一个实验。实验装置的总体质量为 2.608 千克，包括两个相机、控制板、保温材料、数据线、铝合金罐子、18 克水和土壤等。我们用非常简单可行的方式，在月球上构造了一个适合生命生长的、可控的生态环境系统。如果把它放大，下一步就有可能变成一个农场，种植各种各样的蔬菜，变成一个花园，种植各种各样的花草。其实，这是人类在地外星球上生长出来的第一抹绿色，也是这个实验的工程意义，可以为将来载人登月甚至月球生存与月球旅行提供一个探索的基础。所以我们下一步要总结好经验，在这个基础上进一步研究和设计。

信风智库：在设计实验项目的时候，您和团队做了哪些生态系统方面的考量？

谢更新：我们在设计这个实验的时候，肯定也是围绕我们的目标。考虑到人类未来在地外星球的生存，我们从人类必需品的角度选择物种，比如选择了土豆，是因为土豆适合作为太空中的一种主食；比如棉花代表了衣服原料；比如油菜代表了食用油；比如果蝇代表动物，也是广泛用于遗传和演化的实验对象；还有拟南芥，是一种喜光的植物，也被称作植物中的果蝇，是研究中常用的模式生物，非常适合用于植物基因实验；当然还有代表微生物的酵母。我们不仅仅是想让动植物存活和生长，还想让它们形成一个受控的密闭生态系统，植物利用二氧化碳进行光合作用产生氧气，动物利用氧气进行呼吸作用产生二氧化碳，以及微生物的分解作用形成微型生态循环。

因为实验要面向全球发布，我们在选择动物时，还考虑了展示度和宣传效应。我们首先选的动物是乌龟，它有一个壳可以保护自己，在发射升空的时候不会受伤。乌龟也很受大家喜爱，在中国文化里面它是一个代表长寿的吉祥动物。另外，嫦娥三号和嫦娥四号分别携带了"玉兔一号"和"玉兔二号"月球车，我们选择乌龟，也有"龟兔赛跑"小故事的寓意。但是非常遗憾，这次嫦娥四号给这个实验项目的资源非常有限，我们即使让乌龟休眠，它在罐子里待20多天就会把氧气全部用掉。

这个由嫦娥四号搭载的载荷项目的空间是有限制的，即一个直径为173毫米，高为198毫米大小的罐子，除去罐子以及外面的保温层等，里面的生物空间只有 0.82 升。有人问我们为什么不多带纯氧气，这是因为嫦娥四号上的控制信道只留给我们两个，一个用于放水，一个用于拍照，没有更多的控制信道用于释放气体，也没有多余的质量让我们增加其他部件。所以我们只能在 0.82 升的空间里面，填充一个标准大气压的空气，里面含有 22% 的氧气。

后来我们又考虑了用"蚕宝宝"来做动物实验。蚕也很吉祥，代表了我们中国的丝绸文化和丝绸之路，而且蚕用茧把自己包裹起来，在飞船发射时也可以保护自己。然而，蚕非常娇嫩，在背向太阳的低温环境下，很可能会被冻死，所以我们尝试安装含有电池作为能源的保温装置。但是，第三方评估认为电池有可能引发爆炸，所以这个方案最终也被放弃了。

对于这个实验，我们花了很多心思去设计。一是要有科学研究意义，同时还要可行；二是要有现实需要，比如吃穿用等；三是要有中国文化的传承和传播，向青少年、社会大众以及国际社会宣传我们中国的成绩与文化。

所以大家看到，2019 年 1 月 4 日我们在月球上的棉花种子发芽了，一直到 1 月 12 日照相机关机甚至到第五个月，叶子还都是绿色的。这就说明光合作用还在进行，有二氧化碳生成，也说明里面的果蝇也存活着。但是很遗憾，因为资源非常有限，我们的照相机只是一个"傻瓜相机"，只能照到棉花发芽的那一个点，没有拍到其他物种。

仰望星空的民族才有未来

信风智库：深空探测的主要目的是什么？它对我们的国民生产生活有哪些影响和意义？

谢更新：其实无论科技再怎么发展，有两个问题我们必须面对：一个是人到底怎么来的；另一个是我们生活的世界怎么来的，它到底是怎么回事。

古时候，人们不知道闪电打雷是怎么回事，不知道这原来是云层和电离的作用，古人以为是神灵的作用。以前我们不知道空间中还存在无线微波，后来当我们发现无线微波后，就利用它来实现手机无线通信。

所以整个宇宙是非常奥妙的，探索是无穷的。深空探测就是在这种人类探索未知的欲望和好奇心下产生的，把我们认识宇宙的空间维度，一步步地往更深更远的地方拓展。这也就推动我们大开脑洞，想尽办法要登上月球，要遨游太空，要在地球的同步轨道上绕一圈，所以我们发明了火箭、卫星、空间站等，这一过程自然也带动了科技、经济与社会的发展。

深空探测拉动的不只是宇宙空间距离和维度上的更深更远，更多的是我们科技与思想的长足进步。当我们站在太空中看地球，在宇宙之上思考生命，我们的人生观、价值观都会发生变化。就像德国哲学家黑格尔所说，一个民族就需要一些仰望星空的人。如果一个民族没有仰望星空的人，这个民族注定是没有未来的。这就是从哲学上来讨论深空探测。说情怀也好，追求也好，梦想也好，恰恰是这些东西促进社会进步的。而不是说赚了很多钱，就促进社会进步了。

这也是为什么世界各国都非常重视深空探测，它是推动社会和人类进步的重要手段。所以阿姆斯特朗说："我的一小步，却是人类的一大步。"

信风智库： 深空探测和数据收集与运用，有着怎样的关联？数据如何促进这方面的科研？

谢更新： 我国有十几位航天员乘坐过航天器进入太空，我们这些科学家和工程师都没有这样的经历，那么，我们的研究只能基于大量的深空探测科学数据。比如，我们没去过月球，那么怎么设计这次生物科普试验载荷项目的载体罐子？它的造型应该是怎样的？尺寸应该多大？怎么控制材料温度？这些设计完全是基于以前积累的大量数据。所以，深空探测第一个重要成果就是数据。

谁拿到了数据，谁就掌握这个领域的话语权。比如嫦娥四号首次着陆月球背面就拿到了很多数据；比如嫦娥五号登月，拿到了月球土壤，包括月球环境的一些数据。同时，我们也只有得到了这些大量的、第一手的探月数据后，才能对月球展开科研。不然就只能利用外国的数据，不能取得真正意义上的国际领先。这些深空探测的数据采集，包括环境、天文、气象、轨道等方面，都非常有价值。

深空探测的一个重要保障手段就是深空测控网。它在全国乃至全球都布局了数据接收点和控制点，用于与太空上的航天探测器连线，进而获取、发布与传输数据。也是基于这个网络，我们可以通过数字信号来操控太空上的航天探测器。整个过程，数据发挥了第一关键的作用。如果我们的数据不对，算法模型不好，任务就会失败。所以对于深空探测的数据，一方面，我们要有序地管理起来；另一方面，我们要有序地发布，有些数据可以跟全世界共享。

信风智库： 从全球范围来看，在技术、模式与格局上，如今的深空探测与过去相比，发生了哪些变化？

谢更新： 对于科技探索而言，太空环境有着无可比拟的独特价值。比如，太空的辐射和高低温等环境条件可以使基因改性或改良，这是我们在地球上用辐照或紫外线不能比拟的。很多生物公司把一些种子送上太空再返回地球，经过培育与优选，就能获得更加优质的基因改良品种。比如，我们很多科技前沿的研究，需要真空或超低温的条件，在地球上达到这些条件需要很高的成本，甚至难以达到，但在太空就能拥有天然的环境条件。所以，这是世界各国都非常重视深空探测的一大原因。随着我国科技和经济的发展，也早在 2004 年就正式启动了探月工程。

从技术上看，如今的深空探测与 20 世纪 60 年代相比有了很大的变化，一是通信控制更方便、更简单了，设备信道更先进了；二是材料技术的突破使材料本身更轻量化，硬度、强度与防辐射的能力大幅提升；三是计算能力的显著提升，以前一个计算机有房子那么大，现在只需要一部手机就能超过当时的算力。

从模式上看，如今的深空探测和月球探测不像冷战时期，美苏两国为了国家利益与政治较量，展开太空竞赛，而是更多地跟科技进步与产业发展紧密结合起来。每个参与其中的国家，都在思考这个工程如何拉动科技水平的进步；如何让新科技、新技术、新材料与新方法到太空中去验证、去实验；如何让更多的商业公司和民营公司参与进来；如何让深空探测的成果尽快产业化与商业化。

可以看到，美国把很多太空探索的项目交给了商业公司，并且明确了商业公司应该怎么干，要产出怎样的成果。这样一来，这些航天项目产出的成果，就能更快地被应用到经济与社会的方方面面。所以，商业公司发射的火箭次数越来越多，深控探测的频率也越来越高。

而从格局上看，就是国际合作加强。以前只有美苏两国有深空探测的实力。但现在，一是很多国家参加进来了；二是国家之间的合作增加了。前段时间中国与俄罗斯还计划要合作共建国际月球科研站。国际空间站就成了一个国际大舞台，不再是某个国家唱独角戏。全球范围内格局发生变化，形成了一种全球合作的新格局。

信风智库：大数据智能化等新一代信息技术，如何运用到深空探测领域，深空探测又如何拉动大数据智能化技术发展？

谢更新：大数据和人工智能对深空探测的助推作用是非常巨大的。比如我们通过智能化可使探索活动更容易进行，探测数据更容易获得；比如我们可以实现让一台仪器在月球上自动采集数据，并自动把数据传回地球，而且还可以原位就位自动分析这些数据是什么成分，有什么成果。很多从月球传回来的数据，已经经过自动分析形成结果了，更方便地面上的科研人员研究使用，而且数据更精确实时。

目前，在月球探索和深空探测上，我们更多的是发挥智能化的作用，进行无人探索或者无人和有人协同的探索。中国天问一号和美国"毅力"号火星车就是通过大量的自动化与智能化的原位就位探索，然后形成数据包，再传回地球供科研人员进行研究，我们的探月工程也是这样。其实每一台深空探测器就是一个智能化程度相当高的机器人，代替人类从事风险较高的航天探测工作。

反过来，深空探测又为大数据智能化的发展提供了独一无二的应用场景，极大程度地拉动了技术水平的提升和新技术的验证。

一方面，深空探测的应用需求非常明显，任务目标非常清晰，所以它能够准确地牵动技术的提升与实现，来满足需求和达到目标。比如，自主导航和无人驾驶在地球上因为法律法规与伦理道德等问题，暂时还无法投入实际应用；但在月球上和火星上就不会存在这样的情况，探测器都是无人驾驶的。

另一方面，深空探测还能够起到技术验证的作用，在极端特殊的环境中，验证一种新兴技术的可行性和科学性。这就像一个人如果能在极端环境下生存，那么他在普通环境下就更容易适应，生存得更好。深空探测给技术带来了很多极端挑战，就像一个技术的极限运动竞技场，能够对很多前沿技术进行一些前期的验证。

信风智库：请您为我们介绍一下中国探月工程的规划，以及嫦娥号系列探测器的演进过程。

谢更新：对于月球的探索，我国从2004年开始立项，长远规划为"探、登、驻"三大步。探，就是现在正在进行的探月工程，即到月球上探索，摸清楚月球全面的情况。登，就是要有宇航员登上月球，踏一个脚印。这一步还没有实现，我们的天宫空间站就是为这一步服务的。登上月球后，下一步就要建立基地，有宇航员在月球上常驻。

现在我们处于第一大步"探"的阶段，探月工程又分为"绕、落、回"三小步。第一步任务就是"绕"，绕着月球转，是由嫦娥一号和二号来执行的。之前我们通过国际上的文献与报道以及自己的天文观测，知道了月球的地形地貌、环境地理与天文空间等数据与信息，但这毕竟不是我们自己的第一手数据。因此我们就先发射了嫦娥一号，绕着月球转，先不着陆，只是看一看，用光学的、微波的和雷达的方式去扫一扫，嫦娥一号和嫦娥二号就拍下了全月图。

嫦娥一号至六号是两两成对且互为备份的。也就是说，一号和二号的目标和设计是相同的，研制、生产和出厂都是同时进行的，状态也一样。假如嫦娥一号发射不成功的话，嫦娥二号就会马上跟上去。

嫦娥一号成功了，但是嫦娥二号已经研制好了。那怎么办呢？所以嫦娥二号除了拿到全月图，还要把轨道降低。嫦娥一号是围着大圈看月球，因为一号成功了，嫦娥二号就可以做点冒险的事情了。嫦娥一号是在距离月球上千千米的轨道上运行的，嫦娥二号就大幅降低轨道，在 100 千米乘 15 千米这样的一个椭圆轨道围绕月球转。这样就看得更清晰，得到了 7 米分辨率的全月球影像图。

更重要的是，嫦娥二号还为嫦娥三号和四号做了铺垫，把三号、四号的落地点探测好了，特别是得到了落地点虹湾正面 1.5 米分辨率的月球图。所以嫦娥一号、二号获得大量的第一手月球数据为嫦娥三号和四号着陆月球打下坚实基础。

那么,嫦娥三号、四号又需要干什么呢?就是"落",落到月球上去。三号、四号也是互为备份。嫦娥三号成功了,虽然小玉兔不是很听话,只跑了几十米就不动了,但它也拿到了数据,也传回了照片,这也意味着中国成为了全球第三个在月球实现软着陆的国家。

嫦娥三号成功了,嫦娥四号就没有必要再在虹湾着陆了。所以就把四号的目标作了一些调整,把任务难度增大,要四号落到月球背面上去。我们的生物科普试验载荷项目,也是在嫦娥三号成功后增加的,就是要尝试在月球背面做生物实验。嫦娥三号、四号的意义在于让我们拿到了月球上原位就位的第一手数据,而不再是遥测。

嫦娥五号、六号的任务是"回"。"回"不仅是探测器的"回",更是要带回月球的土壤和岩石。嫦娥五号取得了 1731 克月球土壤。我们当时设计的是希望能够拿两千克回来,但是在月球表面钻探的时候,钻得不是很深,钻入的岩层里面土壤太少。对于我们研究来说,月球土壤钻探得越深,研究的价值越大,而月球表面上的土壤数据,我们通过遥感遥测都已经拿到了。苏联通过 3 次探测拿到了 321 克月球土壤,而美国则拿到了 381 千克月球土壤。

作为嫦娥五号的备份,嫦娥六号下一步除了要取样,也要换到月球上的其他地方进行更大范围的探索。这是因为我们需要尽可能地获得更大量、更全面的月球科学数据。特别是嫦娥四号能够到背面去把第一手的就位数据拿回来,这就是我们的贡献。

探月工程的三期规划

期数	任务	探测器	简介
第一期	绕	嫦娥一号	嫦娥一号卫星经地球调相轨道进入地月转移轨道，实现月球捕获后，在200千米圆轨道开展绕月探测
第二期	落	嫦娥二号、三号、四号	二期工程实现月球软着陆和月面巡视勘察等
第三期	回	嫦娥五号、六号	三期工程将实现月面自动采样返回，并开展月球样品地面分析研究

信风智库： 发展深空探测相关产业，对国家和地区有哪些积极的影响和推动作用？

谢更新： 深空探测产业化涉及最广泛的领域就是航空航天。航空航天产业化是现在国家正在规划和重点打造的，比如我国准备出台航天法，航天法里的一个核心就是如何促进我们航天领域有序竞争，如何促进商业化航天与航天技术产业化。

我们在执行航天任务和工程的时候，调动了全国乃至全球的顶尖创新资源。所以，第一个积极影响就是我们航天领域的专家、工程师和管理人员得到了全面的锻炼。这是我们培养人才的一个重要的练兵场，这个作用是不可估量的。为什么现在大家这么重视航天？就是因为航天可以让一些学科和人才队伍成长起来。

第二个积极影响是技术沉淀与转化。航空航天与深空探测领域在技术上是最先进的。我们要用最低的成本、最轻的质量把探测器从地球送到月球，或者把一个庞然大物运送到空间站运行，这里面运用的技术，都是最新最好的技术。那么，这个过程中就会有大量的技术沉淀起来了。

航天领域内的科学家与工程师们，都是为了实现科研目标而全身心地奋斗，比如空间站要在 2022 年实现安全健康运行，大家得想尽办法，怎么控制它，怎么维护它，怎么保证安全，很少去考虑这个技术怎么转化。而我们认为这个汇聚了全国乃至全球科技资源的工程有非常多的潜在价值，值得我们好好挖掘。如果这些商业价值被挖掘出来，便可以引领一些新的产业快速发展，从而造福人类。

比如大数据和人工智能。空间站的整个管理与运营就是全自动化与智能化的，因为宇航员在空间站里不方便干活，不能让他们做很多事情。设计制造空间站的过程就聚焦了大量的智能化科技创新资源，非常值得我们进行商业化的二次开发。

第三个积极影响就是促进产业发展。航天工程是练兵场，不能只练国有企业和国家队伍，还要练广大的商用与民用的技术和产品，所以现在国家也在鼓励让一些民营企业参与到这种重大工程中来。一方面可以培育民营企业的新技术，培养民营企业家敢于担当的精神，另一方面民营企业反过来可以把产业生态做得更健康，发展得更快速。

目前，我们国家的空间基础设施已经初步形成。下一步的深空探测、航空航天、低轨卫星、北斗系统，如何进一步开发利用，如何使更多的老百姓受益，这就需要一大批民营企业的参与。

比如低轨通信卫星涉及的应用太多了，材料制造、通信服务这些产业都要跟上来；比如我们建个空间站，它的拉动作用是非常全面的，值得我们紧跟国家规划去推动产业化。

信风智库：航天领域的民营企业下一步商业化的方向是什么？他们又如何与国家队伍形成一个互补的产业生态？

谢更新：商业航天目前在中国也是刚起步，最重要的是找准自己的定位，在急国家之所急、以国家需求为引领的同时，要跟国家队伍形成互补。

比如现在的民营火箭公司，如果要去做像长征系列火箭那样的产品，就肯定不符合实际，因为没有实力集中这么大量的人财物，也没有长期技术与经验的积累。

那么能做什么呢？可以在专业方面发力，就某个专业或某个点上，商业航天是不是可以干得更深入、更精细化。比如现在有些火箭公司在尝试使用不同的燃料比冲高，将新材料与新燃料作为主攻方向，还有的民营企业就做低成本与轻量化的火箭。所以，商业航天与民营企业要在专业上下功夫，要在细分市场上去找准着力点，而不是做大而全的产品。

再就是要与国家队伍形成互补，能够让国家的航天队伍集中精力在重大任务上，为国家队伍配套。比如像建筑领域，设计院先把房子设计出来，建筑公司就想如何施工建设，至于怎么生产建筑材料，怎么组织施工队伍，都交给外包的合作伙伴了。

民营企业可以做一些专用卫星和以应用型为主的卫星，在研制完成之后，能够马上为市民或产业服务。在生产环节，民营企业在管理上比较灵活，就可以利用新技术，如 3D 打印与一体化成型，这样就能真正地与国家队伍形成配套，也能够在某些专业方面做得更深更远。

信风智库： 因为马斯克的"星链计划"，低轨通信卫星成了目前的一大热点。我们应该怎么看待低轨通信卫星在全球范围内的发展？

谢更新： Space X 公司要发射 42000 个卫星。我们中国其实也在这方面进行布局。低轨卫星肯定是一个趋势，但是它的发展如何，它何去何从，我认为还需要探索。

我们是不是有更好的办法来构建这种低轨通信？低轨通信卫星在太空里的寿命是有限的，那么当这些卫星失去效用后，在太空里怎么清除？这牵涉一系列的问题，目前看来还没有很好的解决办法。

这种太空资源，需要全人类全球共同来商讨一种更先进、更可行、更科学的方案。空间资源肯定是一种非常稀缺的资源。空间资源不像土地资源，中国的领土就是中国的领土，因为地球是自转的，而空间是变化的，是立体的。所以对空间资源的开发和利用，我们的法律和我们的思考都还没跟上技术发展。

下一步要认真思考如何有效地开发利用空间资源。不仅是我们国家要思考，而且整个人类都要思考。不能以跑马圈地的思维去开发，而是要站在人类发展和造福人类的高度、共享共用的高度上去开发。

金贤敏：
量子科技如何打开人类新纪元

金贤敏 上海交通大学长聘教授，博士生导师，区域光纤通信网与新型光通信系统国家重点实验室学术带头人，上海交通大学集成量子信息技术研究中心（IQIT）主任。

2003年考入中国科学技术大学，获得博士学位后，2010年赴英国牛津大学物理系从事博士后研究。2012年同时入选欧盟"玛丽居里学者"和牛津大学"沃弗森学院学者"，并获资助，依托牛津大学独立开展光存储和量子网络的实验研究。2014年受上海交通大学邀请回国组建光子集成与量子信息实验室。2015年获达沃斯世界经济论坛授予的青年科学家奖。

目前已发表论文60余篇。其中包括：以第一作者身份完成的《16千米自由空间量子隐形传态工作》以封面文章发表在 Nature 子刊《自然-光子学》上，并入选两院院士和科技部评选的2010年中国十大科技进展；2018年制备出了世界最大规模三维集成光量子计算集成芯片，并演示了真正空间二维量子行走，相关成果发表于 Science 子刊《科学-进展》，并被评为2018年十大光学产业技术；2020年实现了一种结合了集成芯片、光子概念和非冯诺依曼计算架构的可扩展的光子计算机，相关成果发表在 Science 子刊《科学-进展》上。

扫码观看
访谈精选视频

第一次和第二次量子科技革命，分别为我们带来了怎样的变革？为什么说量子通信可以实现无条件的信息安全保密？发展光量子芯片具备怎样的现实意义，目前研究进展如何？为什么说生物医药、金融交易、航空航天等领域是量子计算绝佳的应用场景？

量子科技带来的改变

信风智库：量子科技的起源是什么？为什么说量子力学是现代物理学的基石之一？

金贤敏：量子科技诞生于20世纪初，科学家们通过对原始理论的创新，把人类对世界的认知从"牛顿力学"转向了"量子力学"，量子科技由此进入了蓬勃发展的时代。

量子力学理论的建立是量子科技最重要的起点。在此之前，我们对世界的认识是"所见即所得"，在万有引力、电磁力、强相互作用力、弱相互作用力这四种力的"大统一理论"形成之后，许多科学家认为物理世界已经晴空万里，除了黑体辐射和以太理论这"两朵乌云"之外，人类基本掌握了宇宙部分真理。

什么是黑体辐射呢？我们知道，任何高于绝对零度（约为 -273.15℃）的物体都会产生辐射，这是一种物体用电磁波把热能向外散发的传导方式。随着热物质温度的升高，电磁波开始不断增强并产生可见光，光的颜色依次变为红色、黄色和蓝色。但是各种颜色电磁波的能量究竟有多少，这个问题一直困扰着科学家们。为了解决这个难题，1900 年量子力学之父普朗克提出"光的能量可以分成不连续的最小基本能量单元"。这意味着一个物理量如果存在最小的不可分割的基本单位，则这个物理量就是量子化的，最小单位就是量子，通过量子就能计算出电磁波的能量值。普朗克的理论，拉开了量子世界的帷幕。

紧接着，爱因斯坦、海森堡、薛定谔和玻尔等科学家，逐渐完善了量子力学的理论框架。在经典物理学中，牛顿力学可以精确地预测物体的位置和速度。1926 年薛定谔提出用波动方程来看待量子世界，即一切粒子在没有被观测之前都是不确定的，它们只是弥漫在空间中的概率而已。在此基础上，玻尔和海森堡提出了"哥本哈根诠释"，根据这个诠释，当对一个系统进行测量时，描写微观世界状态的波函数就会坍缩，粒子的叠加状态也会变成一个位置，这种不确定就会消失。也就是说，是观察者的观测改变了这个系统。

网络上有个段子叫"遇事不决，量子力学"，这句看似玩笑的话却体现了量子力学理论的重要性。它决定了我们对这个世界的认知方式，比如在化学和生物学方面的微观现象，仍然需要依托量子力学理论去作解释。除了在哲学前面不能加"量子"之外，几乎所有的自然科学都有望通过量子力学理论去深层次认识这个物质世界，量子力学已经成为人类认识世界的重要基础之一。

信风智库：第一次和第二次量子科技革命，分别为我们带来了怎样的变革？

金贤敏：20世纪50年代左右，科学家们在研究量子力学规律的过程中，催生了很多现代信息技术，第一次量子科技革命也由此诞生。

第一次量子科技革命最重要的贡献就是发明了半导体。我们知道，半导体是一种常温下导电性能介于导体与绝缘体之间的材料，半导体与金属接触时会出现一道电阻墙，但是电子在运动时却可以穿过这道电阻墙。在经典力学领域，无法解释一个物体穿墙而过的现象，而量子力学却可以通过"隧穿效应"来解释，比如我们常用到的路由器，其无线信号发出的电磁波就可以穿墙。

如果没有基于量子力学对半导体的理解，我们就无法发明硅晶体管以及后来的微芯片、现代计算机和全球定位系统。从某种意义上来讲，量子力学是现代信息技术的理论基础，第一次量子科技革命也直接促进了现代信息技术的蓬勃发展。

这一次量子科技革命对我们这个世界的影响如何呢？有人做过统计，现在70%的国民经济跟量子科技有关，可见量子科技的深远影响。

1989年，随着量子相关实验技术的进步，人类已经可以主动操纵和测量单个量子状态，量子科技迎来了第二次科技革命。以光量子为例，我们可以实现对单光子剥离、操纵和探测，但这个过程其实非常困难，比如一个15瓦的电灯泡每秒钟可以发射出1021个单光子，而我们需要对每一个光子进行信息处理。

光作为量子的信息载体，是一个很好的突破口。我们能够在室温下利用光的波粒二象性、多粒子叠加和纠缠等特点，从

"对量子规律被动的观测和应用"变成"对量子状态的主动调控和操纵",最终掌握了搭建整个量子世界的每一块积木,这样就产生了第二次量子科技革命。

第二次量子科技革命衍生出量子通信、量子计算与量子精密测量三种核心技术。量子通信可以实现近乎无条件的安全通信方式,是通信网络和信息安全领域的一种全新改变;量子计算突破了硅基计算机的物理限制,指数级地提升了计算能力,为 IT 信息产业带来了巨大变革;量子精密测量是基于微观粒子系统及其量子态的精密测量,在测量精度、灵敏度和稳定性等方面大幅超越传统测量技术。

信风智库: 为什么说量子通信可以实现无条件的信息安全保密?

金贤敏: 量子通信是一种信息传递的新型通信方式,通过"量子密钥分发"+"一次一密",实现无条件的信息安全保密。

量子密钥分发就是量子密码,利用了量子不可克隆定理和海森堡测不准原理。量子不可克隆定理,说的是单个量子能量不能再分,一旦操控或测量就会破坏量子能量状态,因此任意量子态无法实现精确复制;海森堡测不准原理即不确定性原理,说的是在微观世界,没办法同时精准地知道量子的位置和速度,位置的不确定性程度越小,则速度的不确定性程度越大,反之亦然。

在这两个理论的基础上,1984 年,IBM 的科学家提出量子密钥"BB84 协议",被称为量子密钥分发的实施方案。那如何实施呢?我们知道,比特是经典计算机中信息的基本单位,而量子比特就是量子信息的基本单位,只是增加了物理原子的量子特性。量子通信的数据载体就是经过偏振编码的量子比特。

由于量子的随机性，我们在传输一段偏振编码由 0 和 1 组成的随机比特时，接收方在接收每个比特时都有一定的出错概率，双方通过经典通信方式沟通一下，删除错误的比特，剩下的比特就是随机产生的正确密码。

如果无授权的第三方截取了量子比特，最后得出的结果往往出入很大，通信双方对比一部分数据发现错误率较高，则证明有窃听者。

另外，一次一密是唯一被证明"绝对安全"的对称加密机制，即要求用于加密的密钥长度与被加密的明文长度相同，要达到这种加密的标准，就需要有很多密钥且密钥安全，这同样增加了第三方的窃听难度。

量子通信的另一个方向是量子隐形传态。单光子在物理属性上不可分割，是两个处于纠缠状态的配对单光子，其中一个无论是左旋还是右旋，都会影响另一个单光子的旋转，它们之间存在一种未知的联系，这种联系的传输速度甚至超过了光速。这就是"量子纠缠"现象。

在量子隐形传态过程中，分别处于遥远两地的甲乙通信双方，首先分享一对纠缠粒子，甲先将待传输量子态的粒子进行分辨，然后将结果告知乙，乙则根据得到的信息进行相应操作。在量子纠缠的"帮助"下，待传输的量子态如同经历了科幻小说中描写的"超时空传输"，在这个过程中，原物始终留在发送者处，被传送的仅仅是原物的量子态。目前，量子隐形传态还处在实验室阶段，离实际应用还非常远。

信风智库：目前，量子通信产业化进程如何？这一过程还有怎样的挑战？

金贤敏：量子通信是量子科技产业落地最早的技术之一。

目前的发展方向是，通过光纤实现城域的量子通信网络，通过中继实现城际的量子通信网络，通过卫星中转实现远距离量子通信。

我国在这一方面发展迅速。2016年发射了全球首颗量子科学实验卫星"墨子号"，2017年建成了世界首条量子保密通信干线——"京沪干线"，2021年成功实现了跨越4600千米的星地量子密钥分发，构建了天地一体化广域量子通信网的雏形。

当然，目前量子通信还存在一些问题，特别是编码和传输方式。比如目前的量子通信终端，物理体积庞大，研制成本高昂，只能应用于金融机构和政府部门，无法普及推广。而实际上，对于工业互联网、智能家居和自动驾驶等诸多领域的通信安全需求，量子通信大有用武之地。

我们也在思考如何真正能做到低成本、便于普及应用的量子通信。其中的关键就是实现量子通信技术的芯片化，将量子通信芯片植入各种小型和微型终端设备上，研制低成本、可移动以及安全可靠的产品与应用，形成与干线级量子通信网络互补的局域网络或支线网络。

基于芯片化这一方向，我们已经实现了量子通信在智能家居、工业互联网等方面的应用试验，其中还包括无人机的影像传输与飞行控制。这些应用场景对通信安全都有很大的需求。这是一个万亿级的市场，我们希望量子通信可以赋能城市物联网和工业互联网，真正走进了千家万户。

信风智库： 为什么量子计算能够带来指数级的算力提升？

金贤敏： 量子计算是第二次量子科技革命中最重要的方向，它能够大幅度提升社会的算力和智能水平，渗透到我们每个人的生活，对整个世界的革新意义要大于量子通信和量子测量。

当然，量子计算的技术原理也可以通俗解释。在经典计算机中，经典比特只有 0 和 1 两种状态，但在量子计算中，量子比特不再是简单的 0 和 1，而是一个展开的二维空间叠加状态。其中一个量子比特展开就是一个二维空间叠加状态，两个量子比特则展开一个四维空间叠加状态，如果有 N 个量子比特，展开的空间叠加状态就是 RN（R 代表空间的维度，N 代表量子比特的数量）。这么多粒子如果实现全互联，就可以形成一个同时叠加或并行的空间，我们称之为"希尔伯特空间"，从而实现对信息计算的处理。

经典计算机一直遵循着摩尔定律，如今面临着物理极限的天花板。但是量子计算的潜力巨大，按照 RN 的公式计算，只需简单加入一个量子比特，就能为计算机的算力带来指数级的增长。这就形成了超越经典计算机计算速度的量子优越性，比如在计算加密算法 RSA 时，由于 RSA 算法采用了密码级别最高的 1024 位二进制密码，即便利用当前最强大的经典计算机，破解 RSA 算法也需要花费 300 多万年的时间，而使用体积大小相同的量子计算机只需要几天时间。

我觉得，量子计算这种革新技术方向，会把人类带到一个新的纪元。但是所有太美好的东西，都需要一个循序渐进的过程，我们要做的是不断进行理论基础和技术应用的研究，不断投入大量资金和人才，在各个应用场景进行验证。

量子科技的产业化之路

信风智库：结合您自身的经历，请为我们梳理一下从前沿探索与课题研究，到技术实现与产业化实践的整个过程。

金贤敏：量子科技研究是一条柳暗花明之路。2003年我在中国科学技术大学开始了量子科技相关知识的学习和实践，其中博士论文工作就是在八达岭长城和河北怀来之间架设16千米长的量子信道系统，实现自由空间量子态的隐形传输。

由于水平大气的尘埃较多，对光的衰减影响较强，造成光子在穿透水平大气时损耗严重。相较而言，垂直大气尘埃较少，对光的衰减影响较弱，光子穿透整个垂直大气层的损耗相当于在水平大气中飞行5～10千米的损耗。

因此，我们利用光子垂直穿透大气层来连接卫星，做了16千米远的超规定数值，作为量子通信卫星的前期项目，不仅完美验证了基于卫星开展量子通信的可行性，还为"京沪干线"和"墨子号"量子科学实验卫星提供了实践依据。

2008年，我继续在团队里做博士后工作，负责搭建基于激光冷却原子的量子纠缠存储实验平台。量子存储是量子通信和量子计算的核心。这个实验非常难，由于光的速度达到每秒30万千米，想要冻住纠缠光子并放到原子里面，十分考验技术和耐心，以至于我在实验室睡了两个月。复杂的科研经历让我能够了解各个技术方向，然后把这些技术结合起来灵活应用。

2010—2014 年，我又到了英国牛津大学物理系进行研究工作，在这期间开始接触量子计算与光量子芯片。2014 年我回国到上海交通大学任教，同时在学校的资助下研究量子通信、量子计算以及光量子芯片。在光量子芯片方面，我们做到了芯片制造流程的全链条化，囊括了设计、晶圆、刻蚀、封装、测试和流片。后来我们逐步将研究成果产业化，在 2021 年创立了国内首家光量子芯片和光量子计算机公司——图灵量子，并且很快就获得了近亿元的天使轮融资，成为国内天使轮融资最高的量子科技公司，开始从光量子芯片底层技术研究向大规模终端应用延展。

当然，任何创新都不是一件一帆风顺的事情。量子计算的创新更是一个"鸡生蛋还是蛋生鸡"的问题，在成果没有完全出来之前，你无法证明这个东西的价值。2016 年，我们在上海交通大学的科研团队就面临过资金紧张的困境，项目一度处于停滞状态。在最困难的时候，是政府在科技创新方面的支持，让我们这个年轻团队逐渐走出了困境。

信风智库：量子计算目前从载体上看，分为光子、电子、原子等；从技术路线上看，分为玻色取样、拓扑、超导、随机线路采样等，在您看来，这些不同形式各有怎样的特点，未来发展趋势以及应用落地的前景如何？

金贤敏：首先，量子计算的载体很多，原则上只要能够携带离散化的能量，这个物理体系就可以成为量子的载体。我们从产业化的角度来看，目前全球有三大主流载体，分别是超导、离子阱和光学。

相较而言，超导和离子阱这两种载体有局限性。由于量子的不稳定性，必须将其放在真空或者超低温环境，不然外界环境会让量子纠缠随着时间逐渐丧失，产生量子退相干效应。而光学更像一个"万金油"，光量子具备不跟环境耦合的特性，可以在室温下运行，不会产生量子退相干效应。如果超导和离子阱想跟别的系统连接构建分布式计算体系，中间的连线也需要利用光学。

测量量子计算机的状态会将不确定的量子状态坍缩成一个确定的单个结果，这就需要全新的算法设计，利用量子干扰和纠缠等特征来获得最终的结果。因此，在量子计算的技术线路上，出现了针对不同物理体系的量子算法，包括玻色取样、拓扑、超导、随机线路采样等。

而光学载体所擅长的算法相对较多。比如玻色取样，我们可以理解成一个量子世界的高尔顿板，小球从上方被扔下，每经过一个钉板都有一半的可能向左走或向右走，当有很多个小球随机掉落时，小球下落位置的分布就会呈现一定的统计规律。但是在计算过程中，我们还需要计算小球的质量、入射角度、表面弧度和硬度，而光学具备不受电磁场影响、传输和转换能量极小、传送过程稳定互不干扰的物理优越性，几乎可以忽略困扰小球的各种因素。

条条大路通罗马。未来，超导、离子阱和光学等不同的量子计算体系，将发挥各自在某些方面的独特优势，形成一种混合的量子计算体系。其中，光学不仅在研究过程中"沿途下蛋"的效果非常明显，还可以作为超导、离子阱等不同体系之间的连接器。

目前，量子计算领域受资本青睐的还是光量子计算，这个领域融资额最高（超过 5 亿美元）的美国硅谷 PsiQuantum 就是一家光量子计算公司。在牛津大学期间，我们团队与布里斯托大学的团队同时在研究光量子芯片的量子计算，也同时在争取英国政府的科研资金支持，最终我们牛津大学团队获得了这笔科研项目资金。后来，布里斯托大学团队的学术带头人就去美国硅谷创办了 PsiQuantum。我相信，在技术路线相似的情况下，依靠中国的科研基础、市场潜力与政策支持，我们图灵量子一定能够与 PsiQuantum 齐头并进。

信风智库： 从产业化的角度，光量子芯片具备怎样的优势？

金贤敏： 光子被研究了很多年，有很多应用场景已经比较成熟，宏观光学也早已是中学教科书上的内容。

但是，在计算应用方面，宏观光学存在着缺陷。第一，宏观光学的系统非常脆弱，把这些光学放大镜和透视镜摆在光学平台上，只要稍微改变角度，结果就不一样；第二，要形成一定的光路体系，需要依靠几百个光学元件去实现，而这些光学元件需要庞大的光学平台承载；第三，宏观光学的计算体系搭建非常依赖科研人员的经验，就像餐厅的口味水平依赖厨师的手艺一样，不能像麦当劳那样标准化、可复制。

如何解决这些问题呢？通过加入量子技术，实现将光子集成于芯片的解决方案，满足可扩展性和标准化的需求，我认为是一条必然之路。

其实，超导和离子阱体系也有芯片，只是这些芯片需要放在真空或者超低温这类跟常温环境充分隔离的条件里面，而且围绕芯片的外设装置常常有数吨重，不像光量子集成芯片可以放在常温环境下，体积更小，移动也更方便。将来，我们希望将光量子集成芯片做到像电子计算机主板一样的大小，在各种应用场景上布置，提升各种应用场景的算力，这就是我们对光量子芯片有信心的地方。

信风智库：发展光量子芯片具备怎样的现实意义，目前研究进展如何？

金贤敏：首先我们要区分光子芯片和光量子芯片的概念：光子芯片运用的是半导体发光技术，产生持续的激光束，驱动其他的硅光子器件；光量子芯片是将光量子线路集成在基片上，进而承载量子信息处理的功能。

另外，光子芯片是一种基于硅基芯片的操控技术，利用激光技术可使光子更广泛地应用于计算机中，也依然存在被光刻机卡脖子的问题。但是这个问题，在光量子芯片领域却不受影响，因为光量子芯片的制程不追求物理极限的提升，仅仅利用14纳米的通用制程就能做到。光量子芯片不是为了解决集成电路芯片的困境，而是一种新的计算构架，是量子新型计算领域一个有前景的发展方向。

不过，光量子芯片并非没有困难，最大的瓶颈就是工艺的精细程度。图灵量子在这个方面走在了前面：一是研发了三维集成光量子芯片，利用飞秒激光在二氧化硅材料上进行立体刻画，在几百飞秒时间内，激光就能完成对材料内部的修饰和改性，制造出任意形状的三维结构和光子线路；二是研发了铌酸锂薄膜光量子芯片，这种新兴芯片具有光谱范围广、调制速度快、集成度高、损耗低等优势，有望替代硅光和传统光调制器。

这两种光子芯片让我们具备了全普段光调控能力，既可以做到三维集成，又可以做到高速度、低损耗的调控，进而推动光量子计算、光学人工智能处理器、光通信和光互联的应用和发展。

我们团队花了很大的代价打造全集成光量子计算系统，打通了光源、处理和探测等研究环节，覆盖了设计、封装和流片等光量子芯片制备的全链条。为什么不像电子芯片那样，采用分工合作的模式，把流片等环节外包呢？因为从全球范围看，目前光量子芯片的研制还处于不断迭代的早期，并不像电子芯片那样发展成熟，产业链各个环节的分工已经非常明确。掌握光量子芯片制备的全链条，更有利于掌握核心技术，并且不断快速升级迭代。

量子计算的优越性

信风智库：操纵量子比特数量是量子计算先进程度的评价标准之一，这将如何影响量子计算未来的升级方向？

金贤敏：目前，谷歌宣称用53位量子比特实现了量子优越性，而未来实用化的量子计算，需要同时操控几十个乃至上百个处于纠缠态的量子比特。在量子计算领域，人类所能操控的量子比特数量决定着量子信息处理技术的高度。但是，我认为操控量子比特数量这种方法存在着不准确性。

第一，操控量子比特数量的容错率低。由于量子比特可以是1和0的任何组合，这就使得如果在创建量子运算时一旦出现小错误，或者耦合到物理系统中存在任何杂散信号，最终都会导致计算出现错误。据相关研究机构2018年的统计，在具有5个或更多量子比特的系统上，量子比特操作的错误率超过百分之几。

第二，操控量子比特数量的难度大。以超导量子比特的操控为例，目前每一个量子比特都要连接几根线来调控和读取，操作量子比特打出的脉冲波形还要极其精确，读取量子比特状态的时候，微波腔内的测量精度还要达到单光子的精度。再加上量子比特退相干问题，只有足够长的相干时间，才能在量子比特的"寿命"中执行更多的量子线路，实现更复杂的量子算法。目前很多团队在尝试用新结构或者新材料来提高相干时间，但这仍然是一根难啃的硬骨头。

因此，如何解决操控量子比特数量的问题，需要找到一个真正合理的方式。在量子编码时，不一定非要编码成量子比特，而是可以编码成多维的量子信息单位，扩大空间的尺度。比如光量子计算，按照前面提到的 RN 公式，可以编码到几十个甚至上百个维度，把量子比特的维度做大，也可以成为衡量量子计算先进程度的标志。

信风智库：量子计算将如何赋能大数据与人工智能领域？

金贤敏：在大数据领域，量子计算可以通过并行计算，实现更加精准、更大动态范围的排序。2020 年，上海交通大学物理与天文学院的团队就展示了大规模量子网页排序的研究成果。

量子网页排序是一个基于量子随机行走的量子算法，相较谷歌、百度和必应等搜索引擎的网页排序，量子网页排序的算法更具优势，比如能够提高排序结果的准确性，减少不同节点排序结果的拥挤，区分网络中的主节点和二级节点。

在人工智能领域，量子计算处理部分的变换矩阵可以应用于人工智能的光学神经网络，形成低功耗、低延时的人工智能判断。我们可以想象一下，集成电路芯片工作时需要做很多逻辑运算，才能形成一个人工智能判断；但在光量子计算里面，光学神经网络训练一旦成功，光只需轻松地从输入端运行到输出端，这个人工智能的判断就已经结束。这是一种静态演算，没有那么多逻辑判断，因此具有快速与低耗的优势，在一些人工智能的边缘终端上，具有很高的应用价值。比如，自动驾驶的车路协同，就可以实现低于百毫秒的快速决策，超过人类反应时间好几个量级。

另外，量子算法在寻找最优解的问题上具备优势。目前机器人发展迟缓的一个原因就是学得太慢，需要经过大量的数据训练，才能找到一个最优解。机器人想要快速获取"智慧"，只要通过量子算法和人工智能的结合，让它在人类社会中迅速学习，在寻找最优解的问题上，只需几个月时间就能超越人类。未来5~10年，我相信在量子计算和光学人工智能的结合领域，一定会有很好的产业前景。

信风智库：为什么说生物医药、金融交易、航空航天等领域是量子计算绝佳的应用场景？

金贤敏：在生物医药领域，一款新药的研发需要耗费漫长的时间和大量的资金，而且具有很大的盲目性和偶然性，临床研究的成功率很低。利用量子计算机快速、并行的数据处理能力，可以准确评估分子、蛋白质和化学物质之间的相互作用，检查药物是否会改善或治愈某些疾病。在新药研制的开发、筛选和优化上，相比经典计算，量子计算能够带来指数级的效率提升。

在金融交易方面，鉴于成千上万的资产具有各种各样的关联性，量子计算有助于投资组合优化以及更有效地鉴别关键欺诈标志。目前，这种量子优化已经成功应用于解决实际金融问题，包括最佳交易路线、最佳套利机会以及信用评分中的最优特征选择，而这些存在变量的数据，经典计算机无法高效解决。

在航空航天领域，美国D-Wave公司是世界上第一家商业化的量子计算公司，2013 年就与 NASA 展开了深度合作，为了应对传统计算机无法应付的复杂而庞大的星际轨道数据，D-Wave 公司利用量子退火算法使其坍缩变慢，保留更长时间的量子叠加态用于量子计算，从而解决星际轨道数据问题。

无论是生物医药、金融交易，还是航空航天，都具有对巨大算力的共性需求，比如面对海量、变量、复杂的数据，经典计算无法做到快速和准确，但量子计算却能够完美地解决这些问题，满足高维度的算力需求。

所以，量子计算在将来很长一段时间，可以在这一类问题上展示更多的优越性，并正面影响这一类问题所联系的国民经济。特别是在精确求解这一优越性方面，量子科技可以提升金融高频交易的精准度，或者为搜索引擎带来指数级加速，让研制药物的速度比以前更快，量子计算将为我们生产生活的方方面面带来巨大提升，将由点到面地在各行各业普及。就像当年的电子计算机，刚出来的时候只能解密码，未来的量子计算也很可能会重塑人类文明。

量子计算机的研制历程

时间	项目	简介
2016 年	IBM	全球首个量子计算机在线平台,搭载 5 位量子处理器
2018 年	本源量子	推出当时国际最先进的 64 位量子虚拟机,打破当时采用经典计算机模拟量子计算机的世界纪录
2019 年	IBM	推出全球首套商用量子计算机 IBM Q System One,这是首台可商用的量子处理器
2019 年	谷歌	发布 53 比特的超导量子芯片,约 200 秒就完成了经典计算机要 1 万年才能完成的任务,实现了"量子霸权"
2020 年	本源量子	自主研发的超导量子计算云平台正式向全球用户开放,该平台基于本源量子自主研发的超导量子计算机——悟源(搭载 6 比特超导量子处理器夸父 KFC6-130)
2020 年	中科大潘建伟、陆朝阳团队	成功构建 76 个光子的量子计算机"九章",200 秒能完成目前世界上最快的超级计算机"富岳"6 亿年的计算量

(信风智库根据网络资料整理)

后记
智能时代的年度印记

以大数据、5G、人工智能与物联网等为代表的新一代信息技术，正在交叉与叠加，同时与实体经济深度融合，由此引发人类社会爆发式变革。在"发展数字经济，推进数字产业化和产业数字化"的国家战略下，重庆市明确提出了建设"智造重镇""智慧名城"的目标，并规划了"芯屏器核网""云联数算用"与"住业游乐购"的实施路径。永久落户重庆，每年一度的智博会，以"智能化：为经济赋能，为生活添彩"为题，已经成为具有国际影响力的重要盛会。

作为伴智博会而生的系列图书，"解码智能时代丛书"自2020年出版以来广受好评。它的文字生动可读、通俗易懂，既总结了智博会的交流成果，又展望了全球智能产业的发展趋势。

为了呼应智能时代发展的新趋势，展示智能产业取得的新成果，同时也为广大群众打造一套了解智能时代、融入智能时代的优秀科普读物，中共重庆市委宣传部决定持续性地出版"解码智能时代丛书"。中共重庆市委常委、宣传部部长张鸣同志对"丛书"进行了全面指导，明确提出：要以国际化的标准，将"解码智能时代丛书"打造为智博会的一张文化名片，以及在智能化领域具有重要影响力的系列读物。

"解码智能时代丛书（2021）"立足于重庆市"智造重镇""智慧名城"建设总体战略，围绕2020线上智博会的交流成果、全球智能产业理论和实践的最新探索组织编写。"丛书"策划内容共3种，其中《解码智能时代2021：从中国国际智能产业博览会瞭望全球智能产业》以图文并茂的形式呈现了2020线上智博会的丰硕成果和智能产业的发展现状及趋势；《解码智能时代2021：来自未来的数智图谱》从"芯屏器核网""云联数算用"与"住业游乐购"的角度，解读了重庆在建设"智造重镇""智慧名城"方面的最新实践；《解码智能时代2021：前沿趋势10人谈》涵盖了10个话题，访谈了10位来自全球理论研究和智能产业领域的代表人物，其中既有院士、教授，也有知名企业家。"丛书"构成了一个立体、多维、丰富的观察体系，在2021智博会召开之际，记录了全球智能产业的新成果，展望了全球智能时代变革的新趋势。同时，为了讲好中国故事，并与全世界的读者分享智能产业领域的中国实践，相关创作内容同时配有英文译本。

"丛书"的组织策划、调研写作及编辑出版是一个庞大的系统工程，整体工作由中共重庆市委宣传部策划组织，并在重庆市经信委和重庆市大数据应用发展管理局等部门的配合下，由信风智库和黄桷树财经两个专业团队创作，重庆邮电大学MTI团队翻译，重庆大学出版社出版。

在整个写作及出版过程中，中共重庆市委宣传部常务副部长曹清尧同志，市委宣传部副部长、市新闻出版局局长李鹏同志对"丛书"的写作和出版工作做了具体安排部署，市委宣传部出版处统筹多个专业团队紧密协同，对"丛书"的策划创意和内容质量进行总体审核把关，推动完成了"丛书"的编写及出版；重庆市大数据应用发展管理局副局长杨帆同志、重庆市经信委总工程师匡建同志对创作工作进行了专业指导；智博会秘书处、重庆市九龙坡区融媒体中心、重庆市两江新区融媒体中心与重庆大数据人工智能创新中心、公共大数据安全技术重点实验室对创作工作进行了大力支持。重庆大学出版社特别邀请智博会秘书处何永红主任、重庆市经信委刘雪梅处长，重庆市大数据应用发展管理局法规标准处杜杰处长以及重庆大学的李珩博士、李秀华博士对"丛书"进行了审读，重庆大学出版社组织了多名资深编辑对书稿进行了字斟句酌的打磨，从而确保内容的科学性、可读性及准确性。

如果站在智能时代历史进程的维度上，我们希望"解码智能时代丛书"能够以年度为单位,记录与展望智能化究竟如何为经济赋能、为生活添彩,记录与展望"数字产业化、产业数字化"的实践过程,记录与展望人类文明史上这场伟大而深刻的变革。这样的记录与展望,这样的智能时代年度印记,是有历史意义的。

谨此,致敬中国国际智能产业博览会,并对所有促成本书立项、提供写作素材、执笔书稿编写与翻译、参与本书审订、帮助本书出版的单位与个人,对接受写作团队采访的专家,致以深深的谢意。

编写组

2021 年 7 月

DECRYPTING THE INTELLIGENT ERA 2021

Interviews with 10 People on Future Technology Trends

Chongqing University Press

Preface

Ask the Era Questions, and Let the Era Answer

From the Stone Age, the agricultural civilization, to the industrial revolution and the Internet era, human history has always been named after an era of evolving productivity. Nowadays, mankind has entered the era of the intelligent interconnection of all things with the ternary integration of "human, machine, and things."

Science and technology innovation, led by a new generation of information technology, has demonstrated the complexity and importance unprecedented in human history. Big data, artificial intelligence, Internet of Things, blockchain, quantum technology, and other emerging technologies not only update and iterate day by day, but also cross, integrate, superimpose and produce fission with each other, which triggers profound changes and exponential progress of human society, economy, and civilization.

What exactly does this kind of transformation and progress look like? Various prophecies about novel technologies circulate widely, some expected, some unbelievable, some blindly optimistic, and some alarmist. So, how do we get a forward-looking perspective and cognitive approach that are both macro and micro, thoughtful and practical, realistic and inferential at this moment?

Tao Xingzhi said, "There are thousands of inventions, but the starting point is a question." Asking the era questions is perhaps the best way to perceive this complex and important age of intelligence. To let the times answer our questions is the original intention for planning *Decrpting the Intelligent Era2021: Interviews with 10 people on Future Technology Trends*.

Achieving this idea is not an easy task.

The first is the richness of topics. We have selected 10 hot topics ranging from big data, artificial intelligence, blockchain, Industrial Internet and digital twins, to quantum technology, deep space exploration and brain-computer interfaces, which run through three dimensions of the intelligence era from basic theoretical research, cutting-edge technology exploration to the technology industry applications. For example, professor Zhou Tao, an expert in big data, showed us how to construct the next generation of the data governance system. Wu Xiaoru, president of iFlytek, shared with us how AI improves people's livelihood such as education and medical care. Li Yuanqing, professor of South China University of Technology, led us to explore the telepathy between the brain and computer. And Xie Gengxin, professor of Chongqing University, expanded our sights onto the moon to see how the Chang'e series of lunar probes realize the three-step objectives, namely "orbiting, landing and sample returning," step by step.

The second is the diversity of interviewees. In an effort to present a more three-dimensional and diversified intelligent era from multiple perspectives, we have interviewed 10 people from Turing Award winners to academicians of the Chinese Academy of Sciences, from China's technology giants to local industry leaders, from academic leaders to industry leaders, from listed companies to entrepreneurial newcomers. For example, Joseph Sfaakis, the Turing Prize winner, explained how AI deals with the complexity of different dimensions. Xu Zongben, the academician of the Chinese Academy of Sciences, elucidated how basic mathematical research supports the advancement of AI. Luo Xiwen, global vice president of France Dassault Systèmes, guided us into the virtual world of digital twins and displayed how to "rehearse" our real production and life countless times. Jin Xianmin, professor of Shanghai Jiaotong University and founder of Turing Quantum, brought us into the micro-world of quantum technology to explore the incredible qualitative changes and advances made by magical quantum in the fields of computing, communication, and measurement.

The third is the relevance of the issues. He Dongdong, CEO of Rootcloud Technology, demonstrated that "Industrial Internet has digitalized, networked and intellectualized the manufacturing industry," which coincides with the profound analysis of the essential characteristics of the intelligent era illustrated by Xu Zongben, as well as the summary of digital twins, "Making the virtual and real world mutually mirror," by Luo Xiwen, the vice president of Dassault. And the point that the massive data are produced as the blockchain enables all things to interconnect, asserted by Xiao Feng, the chairman of Wanxiang, happens to relate with that of Professor Zhou Tao, "the immutable properties of the blockchain making the data more credible," which has also been applied to Rootcloud. Coincidentally, Xiao Feng, Joseph Sfaakis and Luo Xiwen elaborated on the important issue of handling the complexity of the interconnection of everything. There are many more such coincidences and mutual confirmations. This is not our deliberate design. These cutting-edge trends naturally converge and blend as rivers rushing into the sea, forming a huge force that upgrades the energy level of human civilization.

Finally, there is time continuity. To understand the era, one needs to have a long-term perspective. If we take 2021 as a starting point, we hope that with this question-and-answer approach, we can select 10 topics each year and invite 10 experts and entrepreneurs from home and abroad in order to witness the transformation process of the next 10 years in the intelligent era. Exposed to the context of one decade, such questions and answers as how technology iterates, how the industry changes, how to update and how to value the historical importance of the trends. The era is driven by the great men while created by the people. To ask the intelligent era questions is to ask the thinkers, researchers, explorers and practitioners in this era questions. No matter which role they play, they are all integral parts of this great era, and they are all creators who add impetus to this great era. In this sense, their answers are also the answers of the era. Therefore, we also extend a sincere invitation to more domestic and foreign experts, entrepreneurs and start-ups for the next 10 years.

Certainly, we consider both questions and answers based on scientific laws and objective facts instead of the imagined and far-fetched. But whether the question or answer, it is difficult to get rid of the limitations of time and vision, especially in the era of rapid progress of intelligence, furthermore, everything is not absolute. Therefore, the questions and answers are by no means authoritative, but are sincerely intended to arouse our attention, reflection and discussion. For example, Newton said that universal gravitation exists among objects while Einstein asserted that there is no gravitation but the curvature of time and space. Both the two masters, however, are shining in the starry sky of human science.

It makes sense to ask the era questions and listen to its answers. As Georgy Bernard Shaw said, "You see things; you say, 'Why?' But I dream things that never were, and I say 'Why not?'".

Tradewind Think Tank
Cao Yifang
July 2021

CONTENTS
> > > > > >

Xu Zongben:
001/ How Mathematics Makes Artificial Intelligence Smarter

Joseph Sifakis:
027/ How Artificial Intelligence Can Be Trusted

He Dongdong:
047/ How the Industrial Internet Reshapes the Future of Manufacturing

Zhou Tao:
081/ How to Build a New Generation of Data Governance System

Wu Xiaoru:
109/ How Intelligence Facilitates Life

Xiao Feng:
133/ How Blockchain Builds a New Type of Production Relationship

Sylvain Laurent:
161/ How to Promote the Integration of the Digital Space and the Real World

Li Yuanqing:
183/ How Brain−Computer Intelligence Nurtures the "Most Powerful Brain"

Xie Gengxin:
213/ What Does the First Green Leaf on the Moon Mean?

Jin Xianmin:
237/ How Quantum Technology Ushers in a New Era of Mankind

Xu Zongben:

How Mathematics Makes Artificial Intelligence Smarter

In the intelligence era, how does the information space interact with human society and the physical world? And how should we understand digitalization, networking, and intelligence? What other issues need to be addressed for the further development of machine learning and even artificial intelligence? How can artificial intelligence master the learning methodology? What issues should be noted in China's research and application of big data and artificial intelligence?

Mathematics and Artificial Intelligence Complement Each Other

Tradewind Think Tank: Human society, the physical world and information space constitute the ternary world. Nowadays, in the era of intelligence, how does the information space interact with human society and the physical world? How should we understand digitalization, networking and intelligence?

Xu Zongben: We need to have a macro understanding of the world. Human society, the physical world, and virtual space (information space) constitute the three elements of today's world. Human beings are social animals, and each individual will interact with one another. Human society is built by such a huge "network of relationships" And the physical world is the place where human beings live. Therefore, when the human society and the physical world combine and interact with each other, the real world is formed, in which everything has a physical carrier.

Why do we regard human beings as intelligent creatures? Because human beings wish to constantly change and improve the real world, and are willing to devote mass creativity and productivity to it. For example, people invent the mirror to have a clear look at themselves and tidy their physical appearance. Through reflection, the appearance in the real world can be projected into the mirror so as to help people accurately tidy their appearances and correct their behaviors.

In that case, a virtual world corresponding to the real world is formed. Although the objects in the mirror have no physical properties, we can in turn know the real world through it and that is the meaning and value of it. This logic also applies to the intelligent era and digital economy. The image in the mirror can be regarded as data and the world in the mirror is a virtual space. We can reflect and refract our real world through the information space, and thus human society will be greatly developed.

Digitalization, networking and intelligence are the three biggest characteristics of the information space.

Digitalization is the basic way to perceive human society and the physical world. Digitalization refracts or projects the reality in a virtual way. In the era of intelligence, by piecing together and combing data fragments of the real world, one will get to know the nature and logic of the real world. The digital economy is the application of digital methods to present and drive economic development.

Networking is the basic way to connect human society and

the physical world (through the information space). If we want to retrieve the real world through fragmented data, there must be a certain network among information. We need to find the network and establish a channel to connect the information space with the real world, and the channel is called networking.

Intelligence is the way virtual space acts on the physical world and human society. Artificial intelligence means to complete the real behaviors in the virtual space at the same level and difficulty as in the real space.

The era to enrich human cognition and improve the real world with the virtual space is the intelligent Era. Digital China, Network Power, and Smart Society are included in the developmental goals of China, and the internal core is to vigorously promote digitalization, networking, and intelligence.

Tradewind Think Tank: Mathematics is the foundation for the future of artificial intelligence. In what ways is it reflected? And how does artificial intelligence counteract the promotion of mathematical research?

Xu Zongben: Virtual space is the mapping of the real world. But the real world is so big that how can we map it all out? Obviously, "a mirror" is not enough. Immense amounts of data must be pieced together to reconstruct the real world. People are required to find out the structural operation and evolution law of data in virtual space, which is called mathematics.

In fact, early mathematics was just a cognitive language to

describe the objective world with the need of measuring things. In the era of digital economy, mathematics itself has become a tool and technology. Mathematics reflects the real world and is an abstraction that originates from the real world but is higher than the real world. Hence, mathematics and artificial intelligence are essentially consistent in the methodology of handling problems, and they are complementary. As the cornerstone of artificial intelligence, mathematics not only provides new models, algorithms and correctness basis for it, but also supports its potential development, which mainly reflected in the following three aspects.

Mathematics provides a methodology for the evolution of all AI technologies. Without formalization, there is no computerization, and data is the basis of formalization. Mathematics provides a model. It is impossible to improve any technical research and develop perceptual experiments to rational models if it cannot be modeled mathematically. Mathematics also provides a direct language and tools for artificial intelligence.

Artificial intelligence is the simulation of human behaviors or abilities. In a given environment, an intelligent agent can adapt to the environment by interacting with the environment on its own, drawing information from the environment, making self-reflection, and improving its ability to solve problems, which is highly consistent with mathematics. Both artificial intelligence and mathematics enhance cognition and solve problems via obtaining and understanding data. From the perspective of mathematics, machine learning is actually an optimization of function space or parameter space, and the two are essentially parallel to each other.

In the course of the role of artificial intelligence, the environment plays a significant role, which can be not only described by data and information, but also modeled mathematically, in which we can judge the behavior of the intelligent agent and further modify its indicators.

A rule is generated in the environment through mathematical models, under which the intelligent agent can make the optimal decision. People, for instance, will take off their jackets when it is hot. Although such action is a subconscious action, it is actually the final decision made by the human brain after calculation and analysis under the regular model because taking off the jacket is the most efficient way to dissipate heat.

Mathematics not only creates new rule models for artificial intelligence, but also provides algorithms and a basis for correctness. Conversely, the deep learning algorithms of artificial intelligence can also provide good feedback for mathematical research. Mathematical problems such as setting hyperparameter and solving complex partial differential equations can now be handled with the help of AI algorithm models and big data analysis.

To give a simple example, weather forecasting used to be mainly based on natural cycles of astronomy and geography to project weather trends. Now if a region is going to hold the Olympic Games, it needs to know the accurate weather conditions of the region 24 hours in advance, which requires accurate weather forecasting.

But how to achieve accurate weather forecasts? A complicated system of partial differential equations for meteorological elements is essential. Weather stations all over the world need to be established to monitor real-time weather element data, share the data to the local weather data backstage in time, and then analyze the critical data with weather forecast models to figure out the future atmospheric motion state and calculate accurate forecast results.

The difficulty of accurate weather forecasting is that it requires a lot of calculations. And the equation is very complicated because the change of weather is like butterfly effect which will be very different with even a little deviation. In the age without computers, it took some researchers a year to figure out the forecast for the next day, and the actual situation might not be right at all. The emergence of artificial intelligence undoubtedly provided supercomputing power support and made accurate weather forecasts possible.

Tradewind Think Tank: Algorithm is actually a mathematical concept at first, which is a calculation method to solve a certain problem. What are the similarities and differences between traditional algorithms and AI algorithms?

Xu Zongben: Strictly speaking, there is no difference between the two academic terminologies. The so-called algorithm is essentially a standardized logic program to solve certain problems. There are two types of subdivided algorithms. One type of algorithm is the mathematical one, which is more theoretical, principled, and based on mathematical logic, without thinking too much about its implementation and application, also called the authorization algorithm.

Another type is computer algorithm, which usually refers to the mathematical algorithm obtained after optimizing the characteristics of the computer structure, and it is more applicable and achievable. Each computer uses different communication principles and scheduling methods, which requires different operating systems to implement. So computer algorithm is designed to make the computer operate more economically and optimally.

Many big data platforms on the market emphasize the use of AI algorithms. In fact, most of them are computer algorithms that have been optimized. From this perspective, there is no essential difference between mathematical algorithms and computer algorithms.

AI algorithms mainly highlight some of the characteristics such as the large scale of processing data. But it is essentially a mathematical algorithm, involving many abstract partial differential equations, linear equations, and traditional algorithms for Sudoku problem. All algorithms that solve problems through artificial intelligence ultimately end up solving mathematical problems, which is enough for us to understand.

Only by Breaking Through the Foundation of Mathematics Can Artificial Intelligence Learn to Learn

Tradewind Think Tank: From the perspective of mathematics, what issues still need to be solved for further development of machine learning and artificial intelligence?

Xu Zongben: This is a very profound question. Obviously, artificial intelligence had turned its technology inflection point from "unusable" to "usable." It is moving from "usable" to "user-friendly," standing on the eve of moving from manual to automation and on the initial stage toward autonomy. The development of artificial intelligence does not only benefit from the use of background disciplinary knowledge such as linear algebra, statistics and probability theory. The core capabilities of artificial intelligence should be algorithms, computing power, and calculations, which are models represented by deep learning, computing resources represented by supercomputing and data resources represented by big data. It is only based on these three basic capabilities that artificial intelligence has become applicable.

Due to the tremendous data resources, China has always been at the forefront of the world in AI applications, which is our advantage. We have applied artificial intelligence to various fields, explored various models, and achieved great benefits. However, from the perspective of applied research, the current stage of development is far from the true goal of artificial intelligence, which is to build a systematic intelligent agent with the basic logic of artificial intelligence. Just like human beings, the intelligent agent will be able to adapt to the environment, make independent decisions, give automatic responses and take autonomous actions. Mankind will be replaced in some complicated labor to effectively solve the problem of high labor costs. And artificial intelligence will free people from high-risk work, exploration and dangers, actually releasing and developing the labor force. Artificial intelligence is not just limited to primary applications such as facial recognition and online payment. We still have a long way to go and there are still many technical bottlenecks that require major technological changes.

Although current artificial intelligence is also applied to data analysis, it cannot yet think like humans. The following issues remain to be solved.

The first one is to form a new statistical theory. Traditional statistics are based on two concepts. One is the basic assumption of "independently-identical distribution." Simply put, things are independent, but they all have a certain systematic and structural distribution. Based on this, sample surveys could be taken because the taken samples can reflect the distribution of the overall.

The other is probability theory. For example, the probability for a couple to give birth to a boy or a girl is 50%, so is the coin toss. However, the premise is that the number of operations must be infinite to meet the theoretical support of probability theory which is also called the infinity or large sample properties.

Traditional statistics pay more attention to the results presented by the data. But in fact, with limited data, results cannot support a good decision-making plan. But big data and artificial intelligence have changed this thinking pattern into a new one which emphasizes the essence of problems. It is hoped that big data will find subtle indicators from traditional statistical theories, then use artificial intelligence to collect quantities of core data and calculates these indicators so as to draw more instructive conclusions. Therefore, the new statistical theory with big data at its core is not a subversion of traditional statistics, but a supplement and upgrade based on it.

Besides, new algorithms based on big data need to be established. Generally speaking, we compare the application of data analysis to cooking. In the past, there were few data samples. We just mixed the available ingredients like oil, salt and food to make a dish quickly without caring its tastes. However, in the era of big data today, the ingredients and condiments are great in amount and rich in variety that can only be stored in the database, with many misplaced contents. In this case, new ways are in need, and here new algorithms appear.

The essence of these algorithms is mathematical problems

such as solving linear equations, graph calculations, optimization calculations, and high-dimensional integrals. Taking intelligent map navigation as an example, how to go from Xi'an to Beijing? In the past, the resolution of maps was not high, and basic routes could be obtained from ordinary maps. But now with the development of information technology, the resolution of the map is getting higher and higher. It is impossible to cover all the data of the cities and roads between Xi'an and Beijing at one time. It can only provide the road information of some of the cities respectively again and again. So what is the shortest way? How much will it cost? In fact, it solves the basic algorithm problem of graph computing in the distributed graph information environment.

The last one is the issue of deep learning. The core of artificial intelligence is deep learning, but how to interpret the results of deep learning is a problem that needs to be solved. In smart medical care, for example, artificial intelligence uses in-depth learning of the big data of accumulated cases to find out the therapy rules, thereby realizing automatic machine diagnosis. However, if a patient came for medical examination and was told that he had leukemia through a smart machine, but he did not accept the result. How would you explain it? The process of deep learning is extremely complicated, and there is no way to explain it through black-box calculations.

Another example is smart finance. When a customer makes a loan, the risk control system calculates the economic status, social relations and other data models, but the result does not meet the loan requirements, and the customer refuses to accept it. He needs an explanation, while we cannot give one.

The deep algorithm is the most basic feature of artificial intelligence. Although not every application can be explained, if we want to apply artificial intelligence to more fields and scenarios, some decisions have to be explained. In other words, if artificial intelligence wants to develop further, it still needs to solve many basic mathematical problems. (See table)

Ten Basic Mathematical Issues of Artificial Intelligence Summarized by Academician Xu Zongben

1. The statistical foundation of big data	6. The modeling of learning methodology and the learning theory in the function space
2. The basic algorithm of big data	7. The theory of non-convex optimization and efficient algorithms
3. The structure and characteristics of the data space	8. The way to break through the prior assumptions of machine learning
4. The mathematical principles of deep learning	9. The way to achieve automatic machine learning
5. Optimal transportation under unconventional constraints	10. Modeling and analysis of AI system combining knowledge reasoning and data learning

Tradewind Think Tank: Meta-learning is learning to learn, which is one of the important breakthroughs for weak artificial intelligence to move towards strong artificial intelligence. Then what does it mean to equip artificial intelligence with learning methodology? What kind of explorations have we been doing so far, and what problems need we try to solve?

Xu Zongben: This is a very good question. What is the current stage of artificial intelligence? In fact, it is still in the primary stage of completing tasks. What does this mean? It means that if we want to do one project, we have to build a system limited to accomplish this task for artificial intelligence. If we want to carry out other tasks, we need to build another independent system. To put it simply, the current artificial intelligence can only do things without methodology, which we call "weak artificial intelligence."

But obviously, our long-term goal for artificial intelligence is to make it truly simulate humans to adapt to the environment, understand language, and even learn to think independently. The strong artificial intelligence in the advanced stage is the exact goal we pursue. The characteristic of strong artificial intelligence lies in the ability to learn methodology, which means the ability of meta-learning.

And what is super artificial intelligence? It is actually a concept from another dimension. First of all, we need to understand that "super" is actually an extension of ability, not personality intelligence. The essence of technology is the way to extend the functions of human organs. In order to survive better, humans use technology to create abilities that are stronger than the human body itself.

For example, humans cannot run fast, so we invented cars; humans have limited calculations, so we invented computers; in order to be able to go to the sky, we invented airplanes. These things that cannot be done by humans can be accomplished by technology. In that case, from this dimension, super artificial intelligence is neither more intelligent than humans, nor will it replace humans. It is only more prominent than humans in certain specific abilities, and only a description of artificial intelligence.

In fact, the developments of weak artificial intelligence, strong artificial intelligence and super artificial intelligence are entwined. And now intelligence of many machines has surpassed human intelligence. Taking computing efficiency as an example, computers are not only fast and tireless, but also free from the impacts of emotions. Part of the reason why Alpha Go could defeat the Go master is that it is tireless and will not be affected by the results, which is its advantage.

Here I would like to give another example. Autonomous driving requires the car to automatically complete one task after another during the drive. If we put a car in a closed space and design a specific program for it, it will certainly not be a problem for it to realize full-distance autonomous driving. However, if on a natural traffic route, it cannot achieve the set targets because the set program has only one fixed task that is to run the entire course according to the route. While in real roads, many random situations may occur, such as strong winds, sandstorms, lane-changing and overtaking of other vehicles, or the sudden crossing of pedestrians. On those occasions, the task changes.

It is actually the problem of meta-learning. The essence of installing various sensors such as cameras and radars in the car is to collect all kinds of environmental information and deal with the information collected so that the car can make decisions automatically.

In the past, the automated processing of information tasks was separated into parts like denoising, tracking, and light balance. The machine could not automatically switch tasks according to environmental changes. But meta-learning lets the machine master a learning methodology to deal with various problems and enables the machine to learn to process a variety of tasks. Thus, the machine can automatically switch various information processing tasks according to various environmental information, and automatically hammer out new solutions. For example, when an autonomous vehicle detects changes in the surrounding environment, it will make in-time responses, such as slowing down, changing directions, or stopping the engine, and it will even choose the best route to continue.

How Mathematics Empowers Innovative Applications in the Intelligent Industries

Tradewind Think Tank: In fact, the intelligentization of big data also provides a broader application possibilities for mathematics. How does it reflect exactly in daily production and life according to your research and practice?

Xu Zongben: Take a medical application case as an example. CT diagnosis is a typical application of mathematics to model and big data to analyze and diagnose. It is known that the basic principle of CT diagnosis is to scan the parts of the human body that need to be inspected by X-rays, and then convert the scanned light signals into digital signals which are input into a computer for processing and analysis so as to identify the lesions of our internal organs.

Although CT diagnosis is widely used, there are two biggest problems in traditional CT diagnosis. The first one is its relatively strong radiation. In fact, one of the main causes of leukemia is the radiation of CT, so hospitals generally allow patients to have CT diagnosis only once a year. The second one concerns the deployment of CT equipment, mainly in 3A hospitals. The imaging department of these hospitals is overcrowded, and it often takes two or three days to queue. And it is difficult for patients in the vast towns and rural areas to receive CT medical services nearby. Problems concerning inaccessible and expensive medical resources should be addressed.

Then how to solve the two core problems with big data and artificial intelligence?

The first solution depends on the system. Simply put, the scanning and imaging processes of traditional CT diagnosis should be separated. It is known that scanning is to obtain signals from X-rays, and imaging is to process CT films and convert light signals into data and images. If separated, the two processes can be operated separately, which can solve many problems.

We can make CT terminals that specialize in scanning available, affordable and accessible to places, including vast towns and rural areas, as well as outpatient departments in hospitals, so that patients can be scanned expediently and timely. The imaging process is still centrally set in 3A hospitals in central cities. In this way, a further rational allocation of medical resources has been realized, which is also in line with the hierarchical diagnosis and treatment system advocated by the nation.

Then we may wonder how about the intermediate process between scanning and imaging? Traditional CT equipment assembles the two processes together limited by signal transmission. However, with the development of 5G, the efficiency of communication transmission has also been qualitatively improved. It can completely realize the remote synchronous transmission of CT scanning signals in real-time, and the negative film data can also be transmitted to each terminal in real-time, thus constructing a new CT diagnosis process.

The second solution depends on technology. The CT scan model could be reconstructed with a mathematical model to minimize radiation. Traditional CT scans the inside of the human body through X-rays formed by high-energy particles. To get a clear image, it must apply a relatively high voltage and current as well as high X-rays dosage. However, strong X-rays will affect cell function and metabolism when penetrating the human body, thereby triggering the mutation of cancer cells at the critical point. That is the basic principle of radiation.

But this situation will be improved by big data and artificial intelligence, and the core is to calculate the precise dosage of radiation. We can build a mathematical model to identify the image using big data. We just need to scan the key signals of our body with a small dose of X-ray, and then put them into the mathematical model for comparison, and finally, we can analyze the the actual data results, and then achieve precise imaging.

In fact, the mathematical model reshapes the relationship between the original projection and its data. On this basis, CT scans can then reduce the dose of X-rays to one-fifth to one-tenth of the previous dose.

By combining the above two solutions, we will create a brand-new intelligent medical product, which is microdose CT. It has been tested by relevant institutions and gradually become mature, and it will be put into mass production. This product firstly turns the CT into a printer. CT scanning terminal can be equipped in various outpatient clinics, towns and villages. We need only to establish one or two imaging centers in each city to complete the interconnection with the surrounding terminal equipment.

Another typical case is to improve the examination and diagnosis efficiency of magnetic resonance imaging through intelligent big data. It is known that the basic principle of MRI is actually to excite and resonate the hydrogen nuclei in human body placed in specific magnetic field, using radiofrequency pulses. After the radiofrequency pulse is stopped, the hydrogen nuclei in the human body will respond with signals at a specific frequency, which then will be collected by a receiver outside the body and processed by a computer to obtain an image.

People should continuously resonate with the magnetic field in the equipment for MRI. It will take a long time ranging from tens of minutes to one or two hours, so it is particularly painful for children and patients with poor mental states. Therefore, as for MRI, the problem that big data and artificial intelligence can solve more than just reducing radiation, but also shortening time.

Through big data, we can create a fast and accurate algorithm model, which will greatly improve the inspection efficiency of MRI.

Simply speaking, the internal structure of each person is different, so it will take a long time for the traditional magnetic equipment to inspect the inside of the human body and obtain feedback signals. An optimal path for the magnetic inspection can be quickly found in the human body and signal feedback can be obtained in the shortest time with the algorithm model.

But how to comprehend this idea? Just like a beeping pilot sound of a phone call, the optimal communication frequency band need to be found to connect the signal to make a call. We have introduced this principle into the study of improving the efficiency of MRI.

The MRI we designed will take a little time to collect the patient's body data, and then quickly set up an optimal plan for signal acquisition and imaging according to the patient's examination needs through some complex mathematical formulas, and finally get personalized magnetic resonance images quickly. This technology has so far been realized and increased the speed of MRI by 20 to 30 times.

In fact, these two cases not only answered the question of intelligent improvement of medical equipment, but also provided us with greater values. At present, the biggest bottleneck of smart medical care lies in the sharing of medical data. We can collect medical big data more comprehensively and standardly through the intelligent application of big data inherent in medical equipment, and the big data of these cases and diagnoses will also have greater value in future smart medical care.

Tradewind Think Tank: Compared with European countries and the U.S., what issues should China pay attention to in the research and application of big data and artificial intelligence?

Xu Zongben: As for algorithm, just like mathematics, it is not confidential and actually open to the public. In the era of big data, it can be said that algorithm is technology because many applications of information technology are the transformation of intelligent results of big data. At present, the internal logic of various smart products emerging on the market, such as of smart robots and smart homes, is based on AI algorithms which are displayed in the form of hardware externally and become various products.

When combining the algorithm and scenario of big data, a good application product can be developed. In this regard, there are slightly different ideas in China and abroad. In fact, there is not much difference between domestic and foreign levels for the research of mathematical algorithm prototype. Due to our strong ability of logical reasoning, we can achieve valuable breakthroughs in academic research. However, we don't pay enough attention to the improvement of the algorithmic mechanism, and we often realize the problem early but take the action late, which makes us grieved. Sometimes we are too eager to achieve success in academic research without accumulation. For example, when we propose a new algorithm prototype or idea, we are eager to publish papers and show the results, while large foreign companies will wonder how to make the algorithm more refined, whether there are any models and scenarios that can be implemented and how to deepen further research and application when seeing the results. For these foreign companies or institutions, it is not important whether to publish papers or not. They pay more attention to the further research and deepening of technology so as to create higher scientific research and market value.

It is true with some situations in manufacturing. Some manufacturing technology prototypes may be created by us but will be cited by foreign companies or institutions after we release them. On this basis, they will continue to strive for excellence, and finally solidify new ideas, new technologies and new processes into a system or a product. When we saw this system or product again, we even couldn't understand it.

Therefore, as scientists, we are actually very worried about the current situation. The remaining problem for the development of our scientific research and industry is to overcome impetuosity, to do things more solidly, and to take a longer-term view. From this point of view, the logic of certain capitals to seek easy money in business is actually variance with the rules of scientific research. The capital requires rapid incubation, iteration, and project change, while scientific research needs patience and meticulous work.

Tradewind Think Tank: The big data intelligent industry is a systematic and value-based intelligent ecology. From the perspective of national strategy, regional development, and intelligent industry, how should we build this ecology?

Xu Zongben: The research and application of a new generation of information technology can be divided into basic research, technical research and applied research. As for the big data industry chain, it can be divided into data collection, supply, storage, processing, and application. Only when the various links in the chain are organically connected can this industry chain truly form productivity.

So, what are the problems of our current big data industry? In fact, the problem is that the industry chain is weak and the resource allocation in each link is insufficiently reasonable, resulting in overcapacity in some links and under capacity in others, so the industry chain cannot fully output productivity.

Typically speaking, many regions have blindly built data centers, causing overcapacity in the data storage link. There is also a misunderstanding that to develop the big data industry is to build a data center, or that the data center is a very obvious manifestation of industrial value.

In fact, data storage is indeed a very important link in the big data industry chain, which will well support the development of communication operators, technology-based enterprises, and platform-based enterprises. The cost of constructing, operating and maintaining the data center is high. However, companies will enjoy low prices for electricity and factory buildings in centralized construction and operation. Therefore, they do not need to build data centers separately, which will help save costs significantly.

However, from a broader perspective of the entire industry chain, can GDP be generated only by data storage? It's like piling up an entire warehouse with rice, it doesn't make sense if we don't use it for cooking. The data center must serve data operations. Compared with data storage, we have insufficient production capacity in links like data supply and data analysis. We should focus on how to improve these links and put more emphasis on basic research.

In fact, the core of artificial intelligence is algorithms. The key point is how to build an algorithm model to solve the problem based on the existing problems. The algorithm and the underlying mathematical research are invisible and intangible, and the results cannot be explicitly expressed, causing it hard to attract people's attention. However, basic research on this invisible "soft power" is crucial for the robust development of the industry chain as well as the autonomous control of science and technology.

Our country is paying more and more attention to this aspect. As the reform of our scientific and technological innovation system continues to deepen, we, as scientific researchers, have seen a broader prospect. With the strong support of the government, we have established the Pazhou Laboratory in Guangzhou, focusing on research on basic theories and core algorithms of artificial intelligence, software and hardware platforms and key technologies of artificial intelligence, and demonstration applications in key industries of the digital economy. Our scientific researchers have obtained a large degree of autonomy, such as the selection and appointment of talents, the selection and the determination of projects, which has fully stimulated the enthusiasm of our team and motivated researchers to apply their professional knowledge to form a set of basic scientific research innovation models for industrial applications.

Unlike other basic scientific research fields, big data and artificial intelligence require more integration of production, education and research because data and algorithms themselves come from daily production and life, and are designed to solve various practical problems in production and human life. In that case, what we need to pay attention to is how to better transform scientific research results into industrial value, which requires market-oriented and enterprise-based innovation.

At this stage, scientific research innovation in China is mainly implemented by scientific research institutions such as universities and research institutes, but there are very few companies with independent research and development capabilities. The problem is that the innovative results made by scientific research institutions are not likely to meet the actual demands and cannot be transformed to industrial value, which makes enterprises unwilling to "pay for it." Therefore, to achieve industrial innovation of big data and artificial intelligence, we must establish an innovation ecosystem for independent research and development with enterprises as the main body.

Joseph Sifakis:
How Artificial Intelligence Can Be Trusted

How should we understand complexity? What conditions need to be in place and what problems need to be solved for intelligent systems to meet the standard of trustworthiness? What other problems need to be solved for weak AI to move to strong AI? What are the issues that still need to be solved for autonomous driving, both in terms of technology and rules? What challenges will the Internet of Everything bring us, and how should we respond to them?

Unravel the Myth of Complexity

Tradewind Think Tank: In essence, intelligence is to solve the problem of complexity. How should we define or understand complexity?

Joseph Sifakis: The complexity characterizes the difficulty to solve a problem. Humans usually face two types of problems: Those in which we try to understand or even predict the world, and those in which we try to change the world to improve our lives.

Depending on the types and the nature of the problems that we try to solve, there are many different types of complexity.

The most common complexity is computational complexity, which studies the amount of computational resources (usually time and space) required by a computer to solve a problem. As long as the scale of the problem is fixed, we can estimate the time to solve it. But when the problem exhibits growth in size, the computational resources required keep skyrocketing. For example, 1 switch has 2 states, closed and open, 2 switches have 4 states, and 3 switches have 8 states. Then when the number of switches is 30,000 or even 300,000, with each additional switch, complexity will grow exponentially. The same is true in system design. Every time a key factor is added, its complexity will increase exponentially, and the calculation time will continue to increase.

In addition to time, computing space is also a problem we need to solve. There are computationally hard problems that I know how to solve by writing an algorithm. Nonetheless, it is impossible to compute their solution as their execution will need amounts of memory that are larger than the dimensions of the visible universe.

Another type of complexity comes from a lack of predictability, which means uncertainty about the outcome of an experiment or a phenomenon. This is very common in our daily life. For instance, we can predict the weather or the position of an electron within some limits, with some probability.

It is particularly obvious in intelligent systems. Taking self-driving cars as an example, they have to predict possible evolutions of their environment and guess with high probability what can happen and avoid dangerous situations.

Another type of complexity characterizes our poor ability to provide a scientific explanation of the phenomena of the world. We call this type of complexity epistemic complexity (from the Greek word "episteme" which means science). When we need to describe the causal relationship between things, we must find a mathematical model that explains the observed relations between causes and effects. For example, we do not have in physics a "Theory of Everything" that is a unified model that could explain all the phenomena observed from the microcosm to the universe, so quantum mechanics and other physical sciences that explain the microscopic world have emerged.

The emergence of artificial intelligence algorithms such as deep learning was original to solve the epistemic complexity, but it seems that these methods still have great limitations. Both deep learning systems and humans can distinguish between cats and dogs. Humans rely on the differences in the shape, coat color, pupils and other details of cats and dogs, but we cannot explain how exactly deep learning systems work. This "inexplicability" precisely reduces the credibility of artificial intelligence, preventing artificial intelligence from being widely recognized globally.

The fourth type of complexity comes from the limitation of human language expression, which is called linguistic complexity. We have feelings, emotions and preferences, but we do not have the language that is expressive enough to specify them precisely. Subsequently, when we want to build an intelligent system at the same level as a human, we cannot build an emotional and affective system to match it.

The fifth is design complexity. How to understand it? Supposing I am a designer and need to design a system. There are many distributed single-processor systems on the execution platform of this system, as well as mobile systems and self-organizing systems. The more sections a system has at the execution level, the more complicated it is to run, and the more likely it is to have problems. This requires us to streamline the system as much as possible when designing an AI system to avoid problems in the subsequent operation.

Tradewind Think Tank: What is the pathway through which intelligence is applied to the process of solving complex problems? What are the characteristics?

Joseph Sifakis: I should emphasize that there are two different ways to solve problems for both humans and machines. Humans have two types of thinking: slow conscious thinking and fast unconscious thinking.

To complete walking and talking requires the brain to perform complex calculations. But we do not know exactly how the problem is solved. Since our early childhood, we can skillfully move our legs and articulate words, which is an unconscious mind cultivated through long-term external stimulation and training. In other words, the brain can directly respond to unconscious thoughts without "thinking."

The other type of thinking is slow and conscious. It is used to solve complex problems. For example, how to proceed to go from one place to another, how to solve a mathematical problem, how to cook a meal. When we think consciously, we figure out procedures consisting of steps that are logically justified and clarify the tasks that need to be completed in each part.

It is the two systems of thinking fast and slow collaborate to achieve human intelligence.

In order to realize the ultimate vision of artificial intelligence, scientists have also made the same transformation to computers. Traditional computers execute programs written by programmers. And we all know that programs are sequences of instructions that we can understand so as to check and trace subsequently. The neural network designed by imitating the structure of the neural networks of our brain is not the case. Although it is able to solve complex problems, we are unable to explain how exactly they proceed.

Intelligence is characterized by the ability to solve three different types of problems. The first type of problem is the ability to understand the world. That is the ability to treat and analyze sensory information, and understand exactly what happens in our environment. For instance, to understand a text or analyze the movement of objects in the space around us.

The second is to update the knowledge in our minds with sensory information and find all possible consequences by reasoning. For example, if I receive an e-mail inviting me for an annual medical check, I will check my availability and will have to store the date and time of the medical visit in my memory.

The third is overall planning. Layer the realization path to achieve the goal. For the previous example, on the day of my annual medical visit, I will have to plan how I go from my house to the hospital, how long the journey will take, and which items to check first.

In general, the ultimate vision of intelligence is to understand the world in an all-around way, connect the existing knowledge with perceived information, and achieve the goal through hierarchical deployment.

How to Establish Credible Standards for Artificial Intelligence

Tradewind Think Tank: What are the conditions that need to be met and what are the problems that need to be solved for an intelligent system to meet the standard of trustworthiness?

Joseph Sifakis: An important standard is to be able to guarantee that they will behave as expected. In other words, they will not do anything harmful to people or the environment. Of course, not all intelligent systems are critical.

If a game-playing robot or an intelligent assistant does something wrong, this may not be such a big issue. On the contrary, if an autonomous driving system or smart energy system fails, it will cause irreparable losses. The core is that we need to guarantee that critical systems will deliver the expected service and their failure will not cause any loss of lives or money.

For this reason, critical systems must be developed according to standards that specify the type of evidence that their manufacturer should provide to guarantee their trustworthiness.

Consider the example of aircraft. When a civilian aircraft is developed, the manufacturer must prove that the failure rate of the aircraft is less than one part per billion per hour in order to obtain flight approval from the Civil Aviation Authorities. If fails to meet this standard, the aircraft they develop will not be able to obtain flight approval.

Intelligent systems reliability is checked by three types of requirements.

The first type of requirement is the basic security requirements of the system, which is also called the robustness of the technology. Technical robustness requires AI systems to take reliable preventive measures to minimize unintentional and accidental injuries. This requirement also applies to situations where the operating environment of artificial intelligence has potential changes, or situations where other agents interact with the system in an attacking manner.

The second is the security of systems, that is, systems can resist attacks and malicious behaviors. This is the well-known issue of cybersecurity. Like all software systems, artificial intelligence systems should also be equipped with careful defensive measures to prevent attacks from criminals exploiting vulnerabilities, such as hacker attacks. Once the artificial intelligence system is attacked, the data and the behavior of the system may be changed, causing the system to make different decisions or to shut down completely. In addition, artificial intelligence systems should have backup security measures to operate in unexpected situations.

The third is the performance that characterizes how well the system does concerning user demands such as operational capability, throughput, bandwidth, jitter, latency and quality. A system with poor performance may be useless even if it is both safe and secure.

Tradewind Think Tank: We have seen autonomous driving in traffic accidents all over the world from time to time. What are some of the issues that autonomous driving still needs to be addressed, both in terms of technology and rules?

Joseph Sifakis: The emergence of autonomous transportation systems is regarded by the industry as an important milestone towards the era of artificial intelligence. The autonomous vehicle is a very representative case, and the problems it faces fully illustrate the obstacles to the development of artificial intelligence.

Despite the enthusiastic involvement and the massive investment of big tech companies and the auto industry, the optimistic predictions about self-driving cars "being around the corner" turned out to be wrong. At present, autonomous vehicle manufacturers revise their ambitions because of technical problems and the erosion of public trust. Even Elon Musk, CEO of Tesla, admitted, " A major part of real-world AI has to be solved to make unsupervised, generalized full self-driving work."

In my opinion, the current design of autonomous driving

systems is not rigorous enough, and society's general attitude towards it can be summarized as the realism of "accepting its risks" and "greed for its benefits". Automobile companies generally believe that traditional and rigorous design methods have limitations and are not conducive to autonomous driving. Instead, they are more inclined to apply past experience to solve future problems. Some companies are even optimistic that they have mastered the right direction, and that technology implementation is only a matter of time.

From a technological point of view, many problems remain to be solved and it will take probably many decades to reach the fully autonomous car vision. We are transforming from a small, centralized, automated, and small-scale available system to a complex, decentralized, autonomous, and large-scale system. This requires establishing a new scientific and engineering foundation, rather than simply combining the existing results of the past 20 years and focusing only on software such as autonomous computing, adaptive systems, and autonomous agents.

Nonetheless, progress is being constantly made in Advanced Driver-Assistance Systems (ADAS). In modern vehicles, we find an increasing number of features such as ABS, anti-collision systems and lane assist.

Nowadays, some levels of autonomy will be reached under limited and safe environments. For instance, many European countries have begun to develop self-driving trucks. Einride, a Swedish autonomous driving start-up, jointly tested T-Pod commercial autonomous trucks in Sweden with European logistics supply giant, DB Schenker.

A T-Pod truck is 7 meters long and can carry 20 tons of cargo. Technically, it has reached the L4 autopilot, which can realize autonomous driving in specific environments. Once encountering complex traffic environments or poor weather, it will switch to manual driving mode. Of course, to ensure safety, the current public driving range of this truck is 9.5 kilometers, and the intersection with social vehicles is only about 100 meters.

I said that machine-learning techniques cannot be trusted because we cannot understand how they work. The entire system is in a black box, and the operating mode is simple "end-to-end." The existing certification standards require strong evidence that they are reliable and trustworthy. This is not achievable today and it will be hard to achieve in the predictable future due to well-known epistemic complexity limitations.

Thus there are two possible avenues to solve this problem. One is to maintain the strict requirements in certification standards, and this probably will be a blocking factor for AI techniques and consequently for fully autonomous cars.

The other avenue would be that we modify the standards or accept the principle of self-certification adopted by some authorities as what in the U.S. As there is no certification from an independent authority, this approach completely relies on the manufacturer's self-management. It depends on governments' judgments to decide which way to apply.

How to Upgrade the Intelligent System with the Internet of Everything

Tradewind Think Tank: From the perspective of intelligence, what challenges will the Internet of Everything bring to us, and how should we deal with these challenges?

Joseph Sifakis: The Internet of Everything is an ambitious technical vision. The idea is to improve efficiency and predictability: First, allow objects to be sensed or controlled remotely across unified network infrastructure, and second, achieve a more direct integration of the physical world into computer-based systems.

On the whole, the Internet of Everything is mainly composed of two parts with very different aims.

The first part is the Human Internet of Things that is intended to be an improvement of the Internet by providing intelligent services to its users. More specifically, the vision is to move towards what is called a "Semantic Web" developed by Tim Berners Lee in 1998. It is proposed to make the entire Internet a universal information exchange medium by adding semantics that can be understood by computers, that is metadata, to documents on the global network.

Instead of the simple search by keywords, Semantic Web places more emphasis on natural language. For example, I could ask "where is the best place for a vacation on the seashore for a maximal budget of 10,000 RMB in May?".

Subsequently, the web engine would provide an answer list of options satisfying in the best possible manner the criteria formulated in my question. It should be emphasized that despite the great efforts of experts in research and development, this vision of the Semantic Web remains largely unrealized.

The second is the Industrial Internet of Things (IIoT). The core idea is to move from automated to autonomous systems by progressively replacing humans in complex organizations. This is a big challenging step forward. So far we are familiar with the use of automated systems, which can operate devices, appliances even vehicles to a certain extent upon design request.

Autonomous systems are very different. They are intended to work without direct human intervention for long periods. They aim to replace humans and are required to deal with a very complex cyber-physical environment. Such systems have to simultaneously process different goals and partly cooperate with humans. Currently, they are used in intelligent transport systems, smart grids, smart farms, and smart factories.

At present, IIoT mainly faces two problems.

The first is the trustworthiness of networking infrastructures and systems. Most IIoT systems are poor in reliability, and even the basic safety and security are substandard. Today, the increase of connectivity brings a drastic impact on the increase of cyber-attacks. But we have not yet explored how to guarantee the security of large-scale heterogeneous systems.

The second is the security of distributed systems. Under the current state of technology, it is difficult for us to guarantee the security of geographically distributed systems, due to the epistemic and computational complexity. In addition, we need to consider the additional problems posed by the use of artificial intelligence coming from the lack of conclusive evidence that they would behave as expected.

In general, despite the significant progress in all directions including artificial intelligence, 5G, and data analytics, the way towards the Internet of Everything will be longer than anticipated because our capability to develop the required groundbreaking scientific will determine the outcome of this undertaking.

Tradewind Think Tank: To discuss from your research, what do you think AI needs to be addressed as it moves towards strong AI? What problems still need to be solved?

Joseph Sifakis: Firstly, I would like to emphasize the difference between weak and strong AI.

Weak AI aims at solving tasks with complex calculations and high repetitiveness. For example, in specific cases as building a Go-playing system or a higher-order equation calculation system, weak AI must be better than that of humans because the performance of rules and definitions of the system is simple and repetitive. But once the application scenario is changed, the system will fail.

The ambition of strong AI is to approach human intelligence, which is characterized by the ability to manage many potentially conflicting goals with different attributes, including short-term goals and long-term goals. Complex tasks like family, professional and social goals management should be done by predicting the external environment.

Although only with one difference in word, there is a big gap between weak AI and strong AI. For weak AI, the complexity is only computational, while for strong AI all the types of complexity that I mentioned are involved.

In the 1970s and 1980s, strong AI researchers discovered that it is hard to overcome obstacles in realizing cognition and reasoning of artificial intelligence. Many scientists and engineers turned to more practical and engineered weak AI. Obviously, they have achieved fruitful results in these fields: artificial neural networks, support vector machines, and even the simplest linear regression theory which can obtain satisfactory results under the support of a large enough amount of data and calculations. These technologies have supported common functions such as facial recognition and voice recognition today, but even so, they are still far from strong AI.

So, what issues still need to be addressed? I would say there are two that I consider the most important.

The first issue is that machines today cannot surpass humans in "situation awareness." Humans are equipped with what we call "common sense knowledge" which is knowledge accumulated in our mind since our birth. Common sense knowledge is organized in a model of the world that has been built progressively and automatically along with our life. These relationships are gradually established through daily training without a special learning process. For example, given a picture of wood and matches, humans will naturally think of "fire." But this is not easy for machines because they lack this ability of reasoning. So far, the ability of common sense reasoning has plagued artificial intelligence for nearly 50 years.

In 2019, Open AI, a well-known AI research institution, once released the GPT-2 language communication system. This universal language model with 1.5 billion parameters caused a huge sensation and was also "interviewed" by *The Economist*. The sentences generated by this model are so fluent and almost real that Open AI publicly stated that it was worried to risk social security due to its perfect performance so they did not fully disclose the model.

Later, a tester entered in GPT-2 "When you pile matches and wood in the fireplace and then throw a few matches in it, you usually want to...?" If this system is smart enough, the answer would be words like "make a fire," but the truth is an utterly wrong sentence. The ability to reason about the basic knowledge of the world has been one of the insurmountable mountains in the AI field for decades.

For artificial intelligence to reach the same capabilities, a database of common sense knowledge is required. Most common sense knowledge has implicit attributes, making it difficult to express relevant information explicitly. Although early researchers believe that a knowledge base can be constructed by writing down the facts of the real world as the first step to automate common sense reasoning, it is far more difficult to implement than it sounds.

The second is how to teach artificial intelligence to handle "new problems." Swiss botanist Jean Piaget said that "Intelligence is not what you know, but what you do when you don't know." When encountering a new problem, humans know how to build a complete solution, and solve the new problem by drawing inferences from one another.

In recent years, artificial intelligence has explored the transition from a single to a comprehensive system, which requires both module recognition and decision intelligence capabilities. In module recognition, a single AI system has been able to be 90% accurate. However, in decision intelligence, we still need to transform the results of module recognition into real decisions. Such decisions are usually not independent of each other, but require multiple simultaneous decisions, and may even be highly risky. Because of this, the goal of artificial intelligence has also undergone essential changes. It is no longer only studying how to reproduce human intelligence on a single computer, but more importantly, how to build a system to solve the super-large-scale problems in the real world. In short, artificial intelligence is gradually moving from principle research to integrated engineering of artificial intelligence.

Such cognitive change is very different from the traditional goals of artificial intelligence. At present, artificial intelligence is more applied to solving global problems such as transportation, medical care, emergency response, and finance on a global scale, and is no longer confined to primitive forms such as chess.

Three Elements of the Innovation Ecosystem

Tradewind Think Tank: In the era of intelligence, the driving role of technology in innovation is becoming more and more important. How should we build an innovation ecosystem that matches it?

Joseph Sifakis: Innovation is the driving force of the modern economy and the application of scientific research results in developing new products and services. The concept of innovation ecosystems emerged during the last decades of the 20th century from the need to accelerate the production and transfer of knowledge to develop innovative products and services. The innovation ecosystem we are talking about today refers to a significant structure for maintaining innovation. The system is produced by the synergy of three participants, namely big enterprises, academic institutions, and startups.

The first is academic institutions, which represent various research institutions and universities, mainly focusing on cultivating outstanding scientists and engineers and carrying out scientific and innovative research in frontier fields.

The second is research projects, which refer to collaborations between enterprises and academic institutions to address common challenges faced within the field and industry.

The third is startups, which are small and flexible enough to achieve the transfer of research results in optimal conditions.

In innovation ecosystems, we have a remarkable synergy among the three factors as they play specific roles: Large companies provide funding and assert new claims, academic institutions provide basic research skills and knowledge, and startups efficiently implement innovation.

What I want to emphasize is that innovation has no unique secrets or models. It is not only economic giants that can occupy an enviable position or play a leading role. The points are outstanding researchers, breakthrough ideas, and the ability to transform these into high-tech products and services, which are more important than financial and material resources.

An innovative economy requires unremitting efforts and involves a variety of structural changes, which cannot be simply attributed to economic issues. Innovation cannot be achieved only by the introduction of regulations, decrees and regulations, instead should be collaboratively produced by various stakeholders such as enterprises, research institutions, and financial institutions, who will be the main implementors of the proposal.

Here, I want to emphasize the human factor in the process of innovation, which is often overlooked in existing policies and practices. I think it is particularly important to establish measures and incentive mechanisms that link academic research with the real economy. Academic institutions are required to reach a critical scale, and an innovative project must accumulate a large amount of multidisciplinary expertise in order to achieve research results. In addition, the creativity and vitality of researchers should be fully recognized, and favorable conditions should be created to attract the best research talents in the world.

Today, mankind faces many global challenges in the environment, energy, population growth, security, health care, and education, especially in big cities. Innovation and its rational application are critical to overcome these challenges and realize social well-being and prosperity.

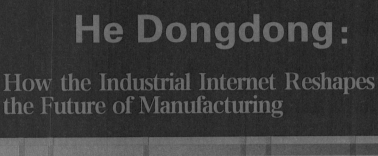

He Dongdong:

How the Industrial Internet Reshapes the Future of Manufacturing

How do we understand the relationship between the Industrial Internet and smart manufacturing? What profound changes will the Industrial Internet bring to the manufacturing industry? What capabilities does a mature Industrial Internet platform need to have? How can the Industrial Internet empower the development of industrial clusters?

Why the Industrial Internet Is a New Type of Infrastructure?

Tradewind Think Tank: How do we understand the relationship between the Industrial Internet and smart manufacturing?

He Dongdong: Smart manufacturing actually describes the ultimate state of manufacturing. When the digital transformation is realized, the manufacturing industry will reach a level, which we call smart manufacturing. Smart manufacturing is a broader concept. From the perspective of the realization path, it actually covers the progress and upgrading of all industrial production factors, including hardware, software, technology, materials, energy consumption, and machines.

The Industrial Internet is the core technology and the most important infrastructure for smart manufacturing. In other words, if the manufacturing industry wants to achieve the ultimate state of smart manufacturing, it must rely on the advancement of industrial Internet technology and rely on the carrying and empowerment of the industrial Internet platform. which is the actual relationship between the Industrial Internet and smart manufacturing.

Why is the Industrial Internet the core technology for smart manufacturing? Whether it is smart manufacturing or Industry 4.0, the core is the promotion of digital technology. Digital technology needs to digitalize the physical objects of industry, and it also needs to integrate digital objects into new generation of information technology. On the one hand, the Industrial Internet has turned various production factors, including man, machine, materials, methods and environment, in the industrial system into various digital objects. On the other hand, it combines new technologies such as big data, artificial intelligence, and cloud computing to drive the automated and smart operation of various production factors that have been digitalized so as to achieve the state of smart manufacturing. This is the technological logic and carrier value of the realization of smart manufacturing through the Industrial Internet.

In addition, from the perspective of stock technology and incremental technology, the Industrial Internet is a necessary supplement to smart manufacturing. The development of global manufacturing industry from Industry 1.0 to Industry 3.0 has actually created and accumulated many manufacturing technologies in informatization and automation. For example, industrial software such as ERP and MES, as well as hardware such as automated production lines and industrial robots, are all stock technologies.

The core of today's Industry 4.0 era is to drive intelligence with digitalization and its specific form is incremental technology represented by big data, artificial intelligence, and the Industrial Internet. However, these incremental technologies cannot subvert the stock technology and exist independently. They must be superimposed on the stock technology and the two can be combined to realize smart manufacturing. Among them, the Industrial Internet is both a technical capability and a carrier platform. From this perspective, the Industrial Internet can also be considered as the last factor to realize smart manufacturing.

Tradewind Think Tank: Looking at the future, what kind of profound changes will the Industrial Internet bring to the manufacturing industry? What are the specific aspects?

He Dongdong: For the manufacturing industry, the Industrial Internet mainly solves three levels of problems, namely, the digitalization, networking, and intelligence of the manufacturing industry.

The first level is digitalization. Why is the digitalization of the manufacturing industry lagging behind the consumer-oriented Internet, which occurred in the early development stage of the Internet? The main reason is that the manufacturing industry is dealing with steel and it still stays in the operation of physical objects. If physical objects are not transformed into digital objects, they cannot be connected to the ever-changing new generation of information technology. The Industrial Internet platform fills this gap by digitalizing physical objects and establishing a "digital twin" or "digital twinship" model, which turns physical assets in manufacturing into digital assets in a real sense.

The second level is networking. When all the relevant physical production factors in the manufacturing industry have become digital ones, the Industrial Internet can connect them all, enabling the interconnection between machines, which can form a larger range of machine collaboration and sharing, so that all industrial production factors are coordinated and operated on the same platform.

The last level is intelligence. On the basis of digitalization and networking, the integration of manufacturing and new-generation information technology has become possible. We apply new technologies of big data, artificial intelligence, cloud computing, and blockchain to optimize the digital model built by the Industrial Internet. In turn, the optimized digital model can better guide us to improve manufacturing efficiency and develop more business models. At this time, smart manufacturing is realized to a certain extent.

In fact, the true value of the Industrial Internet is to bridge the gap between the physical world of the manufacturing industry, the digital world, and the new generation of information technology. It serves as a new type of infrastructure. Why is the Industrial Internet an infrastructure? There are two reasons.

First, infrastructure has a very important feature, that is, it has public attributes and a huge amount of engineering. It is not something that a certain user can build by himself, but a huge public investment, and then everyone can share and use it together.

The Industrial Internet also has this feature. The development of China's manufacturing industry is uneven. Many small and medium-sized manufacturing companies are underdeveloped in informatization and they do not have enough financial resources, materials, and IT technologies to build an Industrial Internet system. Therefore, this requires a more powerful platform or institution, utilizing more powerful technology and resources, to create an Industrial Internet platform for small-and-medium-sized manufacturing enterprises.

Second, after the establishment of the infrastructure, everyone can enjoy the convenience brought by public facilities with a low cost. For example, the construction of infrastructure like high-speed rail requires extremely high investment, but the cost of taking high-speed rail is reasonable and affordable.

Similarly, the Industrial Internet is also inclusive because the Industrial Internet is a platform that concentrates and accumulates a large amount of capital, technology, and resources. It covers almost all the needs of enterprises in the manufacturing field and enterprises can pay on their demand, participate, and use the Industrial Internet in the form of customized services with low access threshold and low cost. For example, in the past, it would cost millions yuan for a traditional manufacturing enterprise to develop an ERP software, while on an Industrial Internet platform, it may only cost tens of thousands of yuan to purchase an interface with the same function and enjoy the latest technology and services at the lowest cost.

The Industrial Internet is like an expressway of information in the manufacturing sector, bringing great convenience of new technologies to manufacturing companies at a very low cost. So we can consider the Industrial Internet as a new type of infrastructure.

Tradewind Think Tank: What capabilities are needed for a truly mature platform of Industrial Internet?

He Dongdong: A mature platform of Industrial Internet must have at least five capabilities.

The first is extensive and reliable device connectivity. On the one hand, it is uiquitous. Our requirement for smart manufacturing is the interconnection of everything. In the future, all sizes of manufacturing enterprises and various machines and equipment will be connected to the platform, covering a number of types, as well as a number of corresponding protocol control instruction set. On the other hand, it is reliable. In a large-scale industrial production system, machines and equipments are running at high speed all the time, so industrial-level stability must be maintained without mistakes.

The second is the processing capacity of industrial data. On the one hand, it must have a strong computing power. The characteristics of industrial data are massive, high-frequency and high-concurrency, and the time unit of data processing is milliseconds. Massive data will come out at the same time within a few milliseconds, and processing, judgment, and decision-making are also required in such a short period of time, which places extremely high requirements on the data link, and streaming computing is also required. This is very similar to automatic driving. The machine must determine what conditions are in the millisecond time unit, and under what conditions it needs to shut down immediately, or even automatically take prevention measures.

On the other hand, it is necessary to have a profound industrial cognition. Whether it is big data, artificial intelligence or cloud computing, if the technology cannot penetrate deeply into the industry and understand the operation mechanism of the manufacturing industry, it will not be able to build a digital model in line with the industrial system and it will not be able to solve the practical problems of the manufacturing industry. For example, when we model the data, it has to match exactly with the industrial mechanism. A machine may be divided into several layers, with a particular architecture, and the internal logic between different modules has to be reflected inside the data model.

The third is open application development capabilities. There are many core needs of manufacturing enterprises. Some need the platform to help with aftermarket management, some need to realize smart production and improve efficiency, and some need the platform to provide business and financial services. Therefore, the Industrial Internet platform must be inclusive so that various industrial application software can open up data and be compatible and integrated with each other. It also needs to be open enough to allow different developers to develop different application software based on the platform for different customer needs.

In fact, the Industrial Internet is also an operating system in the industrial field. For example, the Mobile Internet developers can develop different types of APP applications such as games, shopping and music based on the Android operating system.

The fourth is the accumulation capability for general-purpose applications. The Industrial Internet provides services for manufacturing companies of different types and fields. It can accumulate some general-purpose applications with common requirements and form a plug-and-play service module. Other manufacturing companies do not need to start from zero, which greatly improves the efficiency of application development. Just like accumulating a large number of APPs in the cellphone application stores, the more industrial applications accumulated by the Industrial Internet platform, the lower the cost and faster the enterprises achieve digital transformation, and the greater the value of the platform.

The fifth is the connection ability of ecological partners. The manufacturing industry is a huge ecosystem and no single company can cover all its needs. Therefore, it is necessary to use the platform of the Industrial Internet to integrate the upstream and downstream of the manufacturing industry chain and realize the connection of ecological partners. This is the truly mature Industrial Internet platform.

What Should Be the Path of Digitalization and Intelligence?

Tradewind Think Tank: Data is the foundation of smart manufacturing. In terms of collection, sorting, analysis, application, and governance, what are the differences between industrial big data and the Consumer Internet?

He Dongdong: The Industrial Internet is called as the "third network." In the past, many people interpreted the Industrial Internet as the "Industrial + Consumer" Internet, so some Internet companies signed agreements with manufacturing companies, aiming to construct the Industrial Internet together. However, in the end there is no way to solve the problem. What are the reasons?

At the level of big data, the Industrial Internet is quite different from the Consumer Internet. The first is the difference in the methods of data collection. The Consumer Internet generally collects data through standardized machines such as mobile phones and computers. Although it faces a wide range of users, the commonality of people is the same. For example, collecting information about users' consumption behaviors, what websites they have viewed, and what orders have been placed, which are relatively standardized. However, the Industrial Internet connects thousands of different types of machines and equipments and the mechanisms between the various modules within these machines and equipments and between equipment and equipment are completely different. For example, machine tools and robots are very different and the user behaviors and control logic behind them are different, so the collection of industrial data is generally multi-source heterogeneous and not standardized.

The second is the difference in the amount of data. As the industrial machine runs continuously all the time, it collects data at high frequency and comprehensively for a long time, including process parameters of temperature, pressure, specifications, and energy consumption, which means a huge amount of data. Just like a self-driving car driving continuously, the amount of data generated from acceleration, deceleration, steering and fuel consumption, and other driving commands is very large.

The third is the difference in application focus. Compared with the Consumer Internet, the Industrial Internet pays more attention to real-time collection and rapid decision-making because the core of the Industrial Internet is machine interconnection and the value of industrial data lies in its ability to guide production, predict risks, and improve efficiency. If data processing and analysis cannot be performed through the Industrial Internet in time and the risk of failure cannot be predicted, the consequences will be very serious. Even a small error of one second or even one millimeter will cause a malfunction and lead the machine to stop, which will affect the production efficiency of the entire production line.

It is also related to stream computing. The machine data collection in the industrial system is high-frequency and dynamic. Therefore, the Industrial Internet is required to analyze in real time during the large-scale, constantly changing flow of data movement, capturing potential useful information and send the results to the next computing node to guide the efficient operation of the machine. For example, the cooperation between diesel engine and hydraulic system: The data model must understand how to adjust the instructions of the next link through the instructions of the previous link and control the actions of these machines, which is a very important ability and requires the data model to have a deep understanding of the industrial mechanism — what kind of control instruction a certain data represents, what this control instruction means, and what kind of machine action it will cause, and whether each component or equipment is in a restrictive relationship or a responding relationship in the process flow. Therefore, the Industrial Internet with high-frequency collection, massive data, streaming computing capabilities and an understanding of industrial mechanisms is completely different from the Consumer Internet.

Although both the Industrial Internet and the Consumer Internet are called the Internet and share some common basic technologies, such as storage and computing. However, the logic of constructing data models and the data links used in the Industrial Internet and the Consumer Internet are completely different, which also explains why it is not feasible to use the logic of the Consumer Internet to construct an Industrial Internet as a solution to industrial problems.

Tradewind Think Tank: What are the core issues that need to be solved in the process of the integration of the new generation of information technology and industry?

He Dongdong: Take the development of Traditional Chinese Medicine (TCM) as an example. TCM originated in China and has experienced thousands of years of accumulation and inheritance of "apprenticeship system," such as the research on the efficacy of various herbal medicines and a large number of treatment cases. Since ancient times, the senior the doctor of TCM is, the more skillful he is, which is because the doctor of TCM has accumulated extensive knowledge of pharmacology and cases. But now some changes have taken place. Countries such as Japan and South Korea begin to use mathematics and data to analyze the components of Chinese herbal medicines, and use new technologies to master the mechanism of medicinal materials, forming a new generation of Chinese patent drug system, which has caught up with the development of China.

What problem does this reflect? One follows the new generation of digitalization and the other follows the traditional way of "apprenticeship system" for thousands of years. Obviously, the digital development path is faster and more efficient than the traditional one and it is easier to grow systematically. In fact, on the basis of collecting a large number of folk Chinese medicine prescriptions with mathematical analysis, Nobel Prize winner Tu Youyou discovered that artemisinin can treat malaria.

In the field of industry, the industrial advantages of developed countries in Europe and America are actually very similar to the development of traditional Chinese medicine. They have accumulated a lot of process technologies in Industry 3.0, and many companies were focusing on a certain technology, such as the processing and manufacturing of CNC machine tools. Then based on this technology, the companies continued to optimize their materials, processes and talents, and continued to accumulate their technology. Their process of accumulation and inheritance are similar to the traditional TCM "apprenticeship system" method. Therefore, many European and American manufacturing companies in a certain segment have achieved the world's leading position.

Industry 4.0 has brought a digital development path and provided a new way of technology accumulation. For example, when we need to use a Computer Numerical Control (CNC) machine tool to process a part, the process can actually be expressed by a data model. Specifically, it is to adjust the parameters of the equipment, such as where to cut, what knife to use, how thin and thick to cut, how to adjust the revolving speed. These are the long-term technological accumulations of European and American countries. In the past, even if we get the reverse measurement of the process drawings of European and American manufacturers, or use the same or even more advanced machine tools, but the processed parts in terms of accuracy and quality are not as good as the European and American manufacturers. This is like a veteran doctor of TCM filling prescriptions: The selection of medicinal materials and the amount of each prescription depend on the experience of the doctor.

However, in today's intelligent era, we can build a data model to express the process standards accumulated over a long period of time. Then through repeated production, the results of quality are repeatedly compared with European and American standards. In the process, artificial intelligence is repeatedly trained, the process parameters are continuously optimized, and eventually it will be able to reach or even catch up with the advanced experience level of Europe and the U.S. This is like playing Go with AlphaGo. After the artificial intelligence quickly learns a large number of chess manuals, it can easily defeat human masters.

In fact, this is a difference between the stock technology of Industry 3.0 and the incremental technology of Industry 4.0. Although China's manufacturing industry has not yet reached the stage of Industry 3.0, if we master the new methods of learning and progress paths brought by the new generation of information technology, we can shorten the gap or even catch up with developed countries.

So what are the difficulties in this process? The core is the combination of digital technology and industrial mechanism. Those who understand industry do not understand technology, while those who understand technology do not have a deep understanding of industry. This is also the main challenge of Industrial Internet companies like ROOTCLOUD.

The most direct manifestation is in talents. We need to let Internet talents understand the internal logic and mechanism of manufacturing, then apply the industry cognition to the platform software development. Conversely, for manufacturing-related talents, we must let them understand the new generation of information technology such as data modeling and artificial intelligence, as well as how to build platform tools. In general, it is necessary to combine Internet talents and manufacturing talents so that they can communicate and learn from each other in the process of work and become compound talents.

As for the product, a data model logic that truly conforms to the manufacturing industry should be built and this core technology is called the "object model." Through a relatively unified and universal data model, it can not only model for a variety of machines, but also take into account the inherent mechanism of manufacturing industry, and let software engineers do a variety of applications based on this model. Therefore, in terms of talents and products, the integration of the Internet and industry is the core challenge and the most valuable one.

Tradewind Think Tank: From a market perspective and with practical cases, in your opinion, how to make industrial enterprises, especially small-and-medium-sized industrial enterprises, accept new technologies such as the Industrial Internet?

He Dongdong: Industrial enterprises generally have three attitudes towards digital transformation. The first type of enterprises is generally familiar with the current existing industries, markets, and models, so they are unwilling to take risks to transform and upgrade. The second type of enterprises is afraid to transform. Such companies are more cautious and skeptical of many emerging technologies, thus they are not sure whether the emerging technologies can bring value to the enterprises. The third type of enterprises will not transform. Although they want to transform, they don't know how to do it, and they are not supported by digital talents.

From the perspective of supply and demand, the fundamental reason is lack of money. For manufacturing companies, everyone understands the benefits of automation, digitalization, and industrial software, as well as the principles of digital transformation. However, most of them earn hard-earned money, thus they do not have financial resources to invest in digital transformation. For manufacturing companies, the more they fail to move forward to digital transformation, the more they will become marginalized, and then less money they will have, which forms a vicious circle.

Conversely, for Industrial Internet suppliers like us, the core is to create cost-effective and tangible applications. If you invest 10 yuan, you can earn 15 yuan back. This tangible input and output will be appreciated by manufacturing companies. For a manufacturing company, the questions are very realistic: How to reduce cost, how to sell more equipment, how to reduce the energy consumption by 10%, and how much the company will pay for the solutions. Digital transformation can be achieved as long as the problems are solved.

As a new type of infrastructure, the Industrial Internet has become the most efficient solution. We have invested hundreds of millions of capital and a large number of talents to build such a technology platform. When it is promoted in the SaaS model, manufacturing companies can use new technologies at a low cost.

For example, we have a small-and-medium-sized enterprise customer who makes fruit and vegetable dryers. If they were asked to spend several million yuan to build an intelligent platform from the beginning, they would definitely hesitate or directly refuse. However, they would totally accept if asked to spend only 20,000 yuan to connect 10 dryers at first. During the epidemic, because of the use of the Industrial Internet platform, they maintained several dryers by remotely diagnosing faults online and remotely instructing local electricians to repair them, when there was no way to send personnel on business trips. Previously, people had to be sent on business trips and the response time was slower. Now, not only do we save on travel costs, but we also speed up the response time by guiding local services remotely. Today, this fruit and vegetable dryer company has not only fully used the Industrial Internet but also developed a new business model. In the past, they only sold the dryer to fruit farmers. Now they have developed a mobile fruit and vegetable dryer on the truck. They could directly drive the truck to fruit and vegetable farms in various places, dry the fresh fruits and vegetables on the spot, and then send them to market with high-added value. A manufacturing enterprise that relies on selling equipment has formed a new business model that connects primary, secondary, and tertiary industries by relying on the Industrial Internet.

We can see that the Industrial Internet is a new kind of

infrastructure of centralized investment construction with affordability and production of various personalized application values. In the past, only large enterprises could spend tens of millions of yuan a year to form a team of hundreds of IT engineers to achieve informatization and digital transformation themselves. Nowadays, small-and-medium-sized enterprises can also embrace the intelligent era brought by the Industrial Internet at a low cost. Small-and-medium-sized enterprises constitute at least half of China's manufacturing industry, so the carrier and mode of Industrial Internet are very suitable to enable the new generation of information technology to serve the majority of small-and-medium-sized manufacturing enterprises.

Tradewind Think Tank: What does the establishment of a standard system mean for the development of the Industrial Internet? What are the challenges faced by this work in the implementation process that need to be solved?

He Dongdong: From the perspective of the development of Industrial Internet, establishing standards is the most important thing. We know that one of the key capabilities of the Industrial Internet is the ability to connect a wide range of machines, but one of the challenges is that there are so many different kinds of machines, each with its own different protocols and instructions.

No matter the brand is Huawei, Xiaomi or Apple, we can use the Internet service as soon as we buy the phone because the standards of network access, the communication of mobile phones, and the instruction sets of the three major operators are the same. No matter which manufacturer the mobile phone is, it can be used as soon as the SIM card is plugged in.

I have been advocating for one thing: If we could treat all machines and equipments as mobile phones and network devices, then the standards of communication and the instruction set of networks can be unified, which will significantly reduce the cost of the entire industrial digital services, possibly down to 1% of the original, and will also increase the penetration of the spread of the Industrial Internet significantly. As a platform of Industrial Internet, there is no need for us to manufacture so many communication boxes and invest so much manpower and energy to enable the devices to access the network one by one.

In fact, the manufacturing machine is already a network terminal. If industrial interconnection is fully realized and all machines are connected to the Internet, and then they will all become network terminals. For manufacturers of machinery and equipment, what needs to be unified is the standards of communication and the external instruction set, while the internal components and structures can be kept different, which is the same as the internal components and structures of various mobile phone manufacturers. This will not deprive equipment manufacturers of their differentiated competitiveness.

We then look at the level of the Industrial Internet platform. Any platform cannot be connected to the whole world's equipment, so the manufacturing enterprises and equipment on different platforms need to be enabled with cross-platform collaboration, which requires us to establish cross-platform data exchange standards and data interface standards. Only in this way can the Industrial Internet and smart manufacturing be popularized on a larger scale. Therefore, it is very necessary to establish standards from the two perspectives of widely connected machines and wider applications.

What Changes Will the Industrial Internet Bring to the Industry?

Tradewind Think Tank: At the level of regional economic development, many regions attach importance to the investment promotion of the industrial chain. So how does the Industrial Internet empower the development of industrial clusters?

He Dongdong: Industrial cluster is an important model of China's economic development. We can see that many regions, from the county to the township and even to the village level, gather the same type of business. For example, some regions specialize in making small household appliances and some regions specialize in making textiles. These industrial clusters have two obvious characteristics.

One is that its overall scale is huge but a single enterprise is small. We often see a region where all the companies are doing the same thing and the industry cluster is able to achieve a scale of several billion or even tens of billions of dollars. When it comes to an individual enterprise, its scale is small. This is like a corps of ants, which together can form a very large force, but the strength of a single ant is limited.

Second is that the formats of industrial clusters are generally concentrated in a certain segment of the industrial chain. They either all produce a certain kind of product, or they all sell a certain kind of product, which is the trade link of this product.

What problems do these two characteristics bring? Modern industrial competition is the competition of the industrial chain, including order, production, logistics, and service. Each link requires an enterprise to have advantages which is a manifestation of the comprehensive competitiveness. If they only have an advantage in one section of the industry chain, in today's advanced information and rapidly changing market, they will lag behind in market response. For example, if channels are not opened in time, although the products are good, sales will not be good as expected, and the market value will be greatly reduced. Conversely, no matter how good the sales are, if the quality of the products cannot be guaranteed, the good sales will not last long.

So, how does the Industrial Internet empower industrial clusters?

The first way is to integrate demands and share empowerment. We can integrate the common demands of a certain industrial cluster, empower them through the Industrial Internet platform, and supplement the capabilities that everyone needs. Take the custom-made furniture industry of Guangzhou as an example. We know that custom-made furniture is already a big trend. According to the size and layout of the houses, customers can customize furniture with sizes, shapes and colors that meet their individual demands. So, what kind of capabilities does custom-made furniture industry need? It needs to connect the front-end designer's design software, mid-end manufacturing software, the back-end logistics and installation services in a digital way so as to efficiently and cooperatively promote the process. However, at present, in the custom-made furniture industry cluster in Guangzhou, only a few large enterprises have such ability, and the scale only accounts for 15% of the whole industry cluster. The remaining 85% of SMEs do not have this capability and strength to support a platform. It requires the Industrial Internet to digitalize all links including design, manufacturing, logistics, and installation through a platform-based approach, and then provide it to various SMEs in a method of SaaS. Thus, SMEs can have the same digital customization capabilities as large enterprises for a small service fee.

The second way is to match the demands and respond quickly. Take the small household appliance platform in Zhanjiang as an example. Zhanjiang is very capable of producing rice cookers and hot water boilers, but as the trend of Internet consumption is changing very fast now, when manufacturers finish the links of designing, developing molds, and then manufacturing them, the preferences of consumers may have changed already. Therefore, manufacturers of small household appliances in Zhanjiang need to be more keenly aware of consumer demands and acquire the ability of developing more innovative and fashionable products, as well as the ability of fast molding and rapid manufacturing. The industrial Internet platform we built integrates very high- quality designer resources and fast mold makers, and opens up the whole process from demand to design, to mold opening and manufacturing and sales through information technology and digitalization. Many small home appliance manufacturers in Zhanjiang are able to quickly respond to consumption trends through the platform at low cost.

Tradewind Think Tank: From the standardized mass manufacturing of M2C to the personalized, discrete, and mass manufacturing of C2M, in what aspects does the Industrial Internet provide support?

He Dongdong: To realize C2M, it is necessary to integrate the five links of information, orde, production, logistics, and service so that the entire industrial chain forms a flowing integrity, which includes the synergy of all links upstream and downstream of the whole industry chain such as stores, design, manufacturing, sales, and service. A platform and mechanism are needed in order to connect each link in the industrial chain with each enterprise.

Traditional industrial software can only solve the problem of a certain link. For example, Enterprise Resource Plan (ERP) takes the enterprise as the main service body to complete resource scheduling within the enterprise and the system has internal and external boundaries. On the contrary, the Industrial Internet is inherently an

Internet-based platform mechanism. From the beginning, the entire industrial chain is regarded as a service object, which can break the internal and external boundaries of the enterprise, allowing all enterprises in the industrial chain to coordinate in production on the same platform such as the coordination of orders, design, and production.

There were also some modes of order opening and matching before, such as C2B which only connected consumers with the sales departments of trading companies or manufacturing companies, rather than directly connected with the manufacturing process. Therefore, the order could only be centralized for traditional large-scale and standardized mass production.

In the true sense, the capability of customization of C2M must be reflected in the manufacturing process. Therefore, the "integration of five flows" must thoroughly open up the manufacturing process through the Industrial Internet that connects machines ubiquitously so as to realize the transformation of C2M from consumers to manufacturing.

Tradewind Think Tank: From "selling products" to "selling services," the manufacturing industry has been upgrading in the market end in recent years. In what ways has the Industrial Internet promoted such a transformation?

He Dongdong: From "selling products" to "selling services," this must be a big trend. In essence, when a company buys a machine, its purpose is not to buy the machine, but to buy the service.

For example, a company buys an excavator. It does not mean that the company needs an excavator, but an excavator to dig the soil, which is the ultimate purpose. We buy a car for mobility, but with more smart mobility services available in the future, we will not need to spend hundreds of thousands yuan on purchasing a car.

If we extend the value of the product all the way back, we will find that it is the value of the service in the end. Therefore, the manufacturing industry must also be a service industry. As a result, we have proposed the model of "as a service," such as Platform as a Service (PaaS) and Software as a Service (SaaS). Manufacturing industry can also realize "product as a service" and "operation as a service" through the Industrial Internet, letting customers directly enjoy the services brought by machine products instead of buying machine products.

So in this process, what role does the Industrial Internet play? If the machine is to provide services, it must be remotely controllable and measurable, otherwise the service will not be able to form a transaction. For example, by connecting taxis with the Internet, the Ride-Hailing Taxi platform can measure the miles that the car runs from the start to the end, how much it should charge its passengers, as well as monitor the quality of service throughout the entire process. It is true with the manufacturing industry. The Industrial Internet connects machines and equipments to know how many parts a piece of equipment produces, what the quality of the production is, and who is using the equipment, and then the metering transaction of this production service can be made.

Another important point is the C2M model mentioned earlier.

The Industrial Internet allows the "integration of five links" to open up the manufacturing process and realize flexible manufacturing, providing strong support for the C2M model. In this case, the demand for a personalized service can be directly converted into an order that can be produced and operated in the manufacturing process. Moreover, the intermediate links in process can be eliminated to provide personalized services to the final users more directly.

Therefore, based on the integration of the entire industrial chain, the core capability of the Industrial Internet lies in transforming machines and manufacturing into services, thereby truly promoting the real transformation of manufacturing to the service industry.

Explore the Chinese Logic of Smart Manufacturing

Tradewind Think Tank: In the context of global competition for the development of smart manufacturing, what are the differences between China's smart manufacturing and Europe's Industry 4.0 in terms of development path, mode, and target?

He Dongdong: Let me talk about the similarity first. Whether it is Industry 4.0 or Industrial Internet, it essentially is the digitalization of the manufacturing industry to form data assets, and then to apply a new generation of information technology to the main body of digitalization through "digital twins" or "digital twinship," thus fully optimizing the operating process and business model of the manufacturing industry.

The difference lies in the fact that the foundations and status of industrial development between China and foreign countries. From a global perspective, the development of China's manufacturing industry is large in scale but not powerful in strength. "Large in scale" means that China is currently the country with the most abundant industrial categories, the largest number of industrial machines, and the largest amount of industrial data resources. This is the characteristic of China's manufacturing digital transformation and the advantage of developing smart manufacturing. So why China is "not powerful in strength?" It is because of the uneven development of China's current manufacturing digitalization. Although we have world-leading digital companies like Sany Heavy Industries and Huawei, there are also many small-and-medium-sized manufacturing companies which have hardly achieved informatization and automation. Therefore, from the overall perspective, China's manufacturing industry may not have reached the stage of Industry 3.0 yet, which still needs to be improved.

Based on this status, the development path of China's smart manufacturing will definitely be different from the strategic path of European and American countries.

First of all, the challenges and requirements are different. In most cases, we mainly use the Industrial Internet and digital technology to solve some basic problems such as how to operate and maintain basic equipment through the Industrial Internet and how to accurately allocate labor force and conduct salary administration.

But European and American countries are already using the Industrial Internet for predictive maintenance of aero engines. Although the essence of the technology is the same, the value created will be different if the technology is applied in different scenarios and directions.

Secondly, the imagination space is different. In the era of big data, data is comparable to oil and is the most important production resource, and China has the most industrial machines and industrial manufacturing data, as well as the richest application scenarios. From this perspective, China has great potential for the development of Industrial Internet and smart manufacturing. In fact, the logic of smart manufacturing is that a new generation of information technology is introduced into the application scenarios of the manufacturing industry through digitalization. In this process, China undoubtedly has a leading position in scale.

The last is the different degree of importance attached to smart manufacturing and the Industrial Internet. From the current situation, China's enthusiasm for the Industrial Internet and smart manufacturing is more than that of European and American countries. For many European and American countries with developed manufacturing industries, Industry 4.0 is regarded as a new technological direction and it is enough for everyone to promote it step by step. China has elevated smart manufacturing to a national development strategy, and from all levels of government, to academia, to the business community, all have invested great enthusiasm, attached great importance to it and actively promoted its implementation. Therefore, the future of smart manufacturing in China is very optimistic.

List of Cross-Industry and Cross-Field Industrial Internet Platforms Issued by the Ministry of Industry and Information Technology in 2020

Company Name	Platform Name
Qingdao Haier Co., Ltd.	COSMOPlat Industrial Internet Platform
Aerospace Cloud Network Technology Development Co., Ltd.	Indics Industrial Internet Platform
Beijing Dongfang Guoxin Technology Co., Ltd.	Cloudip Industrial Internet Platform
Jiangsu Xugong Information Technology Co., Ltd.	Hanyun Industrial Internet Platform
Root Internet Technology Co., Ltd.	RootCloud Industrial Internet Platform
UFIDA Network Technology Co., Ltd.	Smart Industrial Internet Platform
Alibaba Cloud Computing Co., Ltd.	Sup ET Industrial Internet Platform
Inspur Cloud Information Technology Co., Ltd.	Inspur Yunzhou Industrial Internet Platform
Huawei Technologies Co., Ltd	FusionPlant Industrial Internet Platform
Foxconn Industrial Internet Co., Ltd.	Fii Cloud Industrial Internet Platform
Shenzhen Tencent Computer System Co., Ltd.	WeMake Industrial Internet Platform
Chongqing Humi Network Technology Co., Ltd.	H-IIP Industrial Internet Platform
Shanghai Baosight Software Co., Ltd.	xIn3Plat Industrial Internet Platform
Zhejiang Lanzhuo Industrial Internet Information Technology Co., Ltd.	supOS industrial operating system
Ziguang Cloud Engine Technology (Suzhou) Co., Ltd.	UNIPower Industrial Internet Platform

Tradewind Think Tank: There is a saying that smart manufacturing is to replace manpower with machines. Based on the continuous evolution of industrial manufacturing, how should we understand the relationship between intelligence and workers from the perspective of the upgrading of actual process?

He Dongdong: There were no machines 1,000 years ago and almost all work relied on manpower. Today, a large number of machines have replaced manpower in many fields. For example, agriculture used to rely on the cultivation by human, but now various machines can be used to cultivate, sow, and harvest, greatly reducing human labor. However, in fact, the employed population around the world is growing on a large scale and has never been negatively affected by the appearance of machines.

Why? We can explain it from two perspectives. The first is that new demands will create new employment. The growth of human demand will always exceed the development of productivity, which is also the source power for the progress of human civilization. For the goods and services provided by the society, people will always have a variety of new needs, so this will give birth to a variety of new productivity and new employment.

For example, the Industrial Internet needs to connect all machines and equipment, so how to connect them stably, how to get through machine data after connection, and how to monitor and use it, which requires a large number of new talents who understand both industry and IT. Therefore, we have specially developed training courses and cooperated with vocational colleges to cultivate such talents, creating a large number of new types of work in this process.

The second reason is that the value of human labor is constantly shifting. For example, in the farming era, almost all labor was concentrated in agriculture. When the industrial era created farming machines, labor force was transferred from agriculture to industry. Although the labor force in agriculture has shifted, the output of agriculture has risen instead of falling because machines have improved production efficiency. Now we have entered the intelligent era. In the process of digital transformation and smart manufacturing, machines will replace traditional production workers. At this time, human labor will shift to digital and intelligent fields such as industrial data analysts, AI engineers, and information security engineers.

Recently, in the fields of global industry and energy, we have seen frequent incidents of cybersecurity, causing immeasurable losses, which means that the intelligent era has created a lot of needs of information security. In the past, a traditional factory might require 20 security guards, but now a smart factory may require 20 network security guards, who are called as information security engineers.

Therefore, from the perspective of the development of human history, science, and technology have never reduced the employed population, but only increased the population and the number of jobs.

Zhou Tao:

How to Build a New Generation of Data Governance System

What is the level and stage of development in the field of big data? How is big data supporting the development of new technologies such as artificial intelligence and blockchain? What issues still need to be addressed for the further development of the big data industry? How to set the standard of data and how to control the quality of data? How to realize data valuation and data capitalization? What basic conditions are needed for the market-oriented circulation of data elements? How should we deal with data governance and data security issues?

How to Activate Amounts of Dormant Data Resources

Tradewind Think Tank: The concept of big data has been born for more than 10 years and it has been gradually accepted. Its various applications have been widely used. From a long-term perspective, what level and stage do you think the development of big data is?

Zhou Tao: It has been nearly 13 years since IBM put forward the concept of big data in 2009. Generally speaking, big data is still in the early stage of development as data has begun to play a role in certain vertical industries and generate value. Actually, more than 90% of the data has not given full play to its value.

A large amounts of data resources are still in a dormant state. The work of identifying valuable data and its application scenarios, and realizing the value of data through a more automated and lower-cost method to control data quality has not yet started on a large scale.

Big data is generally still in the stage of being recognized, knowing that it is a valuable thing. As things stand, many people are still reluctant to give their data to others so easily. But in fact, even if the data is given to others, it is likely that others will not be able to dig up any value, and the data will remain in their own hands, which may also have little value. Big data is still in such an early stage.

Tradewind Think Tank: In your opinion, what misconceptions still exist about big data, whether from government managers, practitioners, or the public?

Zhou Tao: So far, people still have some common misconceptions about big data. The first typical misconception is that data should be large enough to produce value. In fact, the big data mainly refers to the great value of data, which does not mean that we must have a petabyte of data to be called big data. If a very small amount of data plays a supporting role in key decisions, it is also called big data.

The second misconception is that all data must be valuable. In a sense, it may also be right that we first blindly invest lots of cost to store data. Although we do not realize the value of these data at the moment, we may realize it in 10 or 20 years. Yet a significant portion of the data will be of little significant value for a very long time. Then, collecting these data is actually a useless investment.

Therefore, we must distinguish clearly which data is high-value and tradable, which may be of lower value, and which cannot see value for a long time. We need to realize that not all data are resources or even assets.

The third major misconception is that many people believe that after having a large amount of data, they can make predictions and judgments through correlation analysis of data, without relying on industry knowledge and causality. This is actually not true because there will not be a universal data model now or in the future, let alone to use one model to solve all industry problems. Therefore, the cognition of industry is very important, and in most places where we need to intervene and regulate, we must know the cause and effect. If we know the co-relation, we can make some predictions, but cannot make effective interventions.

The fourth typical misconception is to confuse the core functions of big data and artificial intelligence. It is believed that all big data companies are also artificial intelligence companies, and vice versa, which is not true. A considerable part of artificial intelligence depends on data, while much artificial intelligence is not data-dependent, which copes with problems by reinforcement learning, self-learning, or expert decision-making systems. Unlike artificial intelligence, the core capability of big data is to process massive amounts of data efficiently. Through massive data, we can know exactly how many resources and where these resources are located. To a large extent, the better we know, the more precise decisions we could make.

Taking education as an example, what we may need to know is how many different types of educational equipment and how many teachers there are in a province, and what is in shortage in each county. With this data, we can directly make decisions, and it does not need to use complex artificial intelligence. Therefore, I think everyone is confused about this boundary and the concepts of cloud computing, big data, artificial intelligence and even blockchain. In fact, each has its own core competence. If we don't understand the core competence, we are more likely to make some mistakes.

Of course, the last misconception is that many people are overly afraid of the disadvantages that brought by big data, such as invasion of privacy. In fact, very few people have suffered tremendous harm due to the leakage of private data. While there are many people who are overly optimistic and feel as if everything will be data-driven in the future. I think these are more extreme ideas. We really need to pay attention to privacy issues and data ethics. But it is unnecessary to turn pale at the mere mention of these issues because not many people are actually hurt by these problems. We should take a more relaxed attitude to realize that there are good and bad aspects in the development of data industry and data technology, but we cannot unilaterally overstate the good and the bad to a particularly large extent.

Tradewind Think Tank: As a means of production in the intelligent era, how does big data support the development of new technologies such as artificial intelligence and blockchain?

Zhou Tao: You mentioned support. In fact, your question is the answer. To a large extent, big data and artificial intelligence are the mutually supportive relationship. One of the more active branches of artificial intelligence is deep learning. Deep neural networks in deep learning have been proposed for decades. Why has it been so popular in the past 10 years? To a large extent, it is because ImageNet has opened a large number of images with annotated data, so that we can use more parameters in a larger space for some learning without knowing any causality, which needs to be supported by massive data. Therefore, a considerable part of artificial intelligence relies on labeled high-quality knowledge data.

Artificial intelligence is the production tool, and big data is the production material, while blockchain is different. In fact, the blockchain can build and operate without relying on big data at all, and it can build a set of systems by itself. However, the blockchain can empower big data. As with the encryption tool of the blockchain, we can collect more credible data. It becomes more difficult, or even technically impossible for us to tamper and modify the data.

Since the cost of the blockchain is very high, it is not necessarily needed anywhere. The blockchain does not rely on big data, but if there is no data on the blockchain, it will not be valuable. If you compare big data to water, the blockchain is like a water pipe. With this blockchain water pipe, the data will flow more smoothly, otherwise there will be a large amount of data being altered at will and the authenticity can not be judged.

Tradewind Think Tank: In fact, we have seen many big data platforms and systems, which are still in the stage of information. What problems still need to be solved for the further development of the big data industry?

Zhou Tao: Although there are still some areas that can be worked on to improve the database, data storage, computing power and the allocation of computing resources, there are not many bottlenecks that restrict our development for big data. Facing the next development of big data, what is more important is the change of our cognition and management thinking. For example, how the government solves the security and privacy issues of data openness and sharing, and how to evaluate the value of data provided by various data providers in the process of data use and form a positive, perceivable, and recordable feedback, which need to be solved.

In addition, it is necessary to explore how to use some new methods to make our system construction not just stay on the basis of information and business functions. For example, from the perspective of data quality, through data supervision and data quality control, business systems can generate high-value and circulating data. But there are still a lot of problems, including the data quality control and other technical problems.

Therefore, we need to make a lot of efforts to break through the bottlenecks of these concepts and ideas, which is even more important than making technological breakthroughs. We all see now that the government has issued a document that data should be used as a new element of production. But in fact, after the data is generated, it can achieve greater value through multiple circulations, however, it is not yet explored successfully.

Big Data Needs "Three Highs": High Standards, High Quality, and High Value

Tradewind Think Tank: Standards and quality are the two cores of data governance. So how do we set data standards to control the quality of data in the actual application process?

Zhou Tao: The second question is relatively simple. For the quality, people have some accepted standards such as the miss rate of data. To give a simple example, In many cases our data is basically a big table, so whether this table can be filled and how full it can be is the question of miss rate. Another example is that whether the real-time performance of the data is very good and whether it will be a long latency. Is the logic of the data self-consistent? Will it appear in the data that two people are studying in the same school? While the data show that the two people are in different city, the logic breaks down.

Whether the data has some weird and exceptional error values. For example, if there are only 9 digits in the mobile phone number, isn't it an obvious exceptional error? In addition, whether there are many conflicts among the annotations of the data, and whether the annotations are credible. For example, the Chinese written by different annotators in the same voice dialect is very different, and the credibility will be reduced. All these require a holistic approach to describe the quality of the data. For more than 99% of the common data, it is possible for computer programs to automatically judge the quality. Data quality control can be achieved but is not widely used because we still don't pay enough attention to data quality.

The first problem is very challenging. The data standard seems simple, because we always have a standard, which is not only the promotion, but also the restraint to the industry. If there is no standard in an industry, everyone does their own way. It is an obvious constraint that different systems cannot be analyzed together in the future. But once there is a standard, there must be a threshold. The standard will give a boundary, which makes those that are not within the standard, but better things cannot enter. Therefore, standards are also the constraints of the industry.

At this time, we must pay attention when making new data standards. First, we may have to start with needs which means beginning with the end in mind. We will figure out what the demands of this industry are and what problems we want to solve when we set standards, but we cannot set standards with blind action of divorcing ourselves from reality.

The second is that the standards must be flexible. Now when we establish a standard, we may have to leave a lot of fields. For example, it may have been 12 bits. Now we may have to leave a few blank fields on the basis of 12 bits. In the future, as long as the previous basic fields meet the standards, we can use them for anonymization and benchmarking with the same ID. Many previous standards write so completely that it would be very troublesome to add future new things. Therefore, building data standards is more difficult, but generally speaking, it is easily promoted in some more mature industries with data building.

Tradewind Think Tank: Based on normalized standards of data governance and data quality, how to realize the value of data and the capitalization of data?

Zhou Tao: The core of data to realize the value lies in whether there are scenarios, only data with scenarios can realize value. For example, we can rely on technology to obtain the price and know the sales volume of a large number of clothing on the e-commerce platform, and it is likely to be able to predict what colors and styles of clothing will be popular in the season. It's interesting but of little value to us because we don't have a clothing factory. And if we cooperate with clothing factories, this data can guide production of goods. Therefore, only scenarios can make data generate applications and value. Creating or finding scenarios suitable for data applications is the best way to make data available and valuable.

In fact, China has many scenarios, but one thing is still missing, that is to pay for knowledge. For example, for a big data application in a certain scenario, it is not easy to promote sales because it is non-physical software, but when we install this big data application on a computer, it is relatively easy to sell because it is hardware. In fact, people can install this application on any computer. Therefore, it is not easy to make everyone willing to pay for the governance process of data value realization.

Data asset or capitalization is more complicated, which means adding the data into the accounting table, the balance sheet of the company, making pledges, mortgages, and even investing as equity in the future, which is apparently a more difficult task. Data asset requires a third-party appraisal agency that can evaluate the price or value of data so as to promote data asset or capitalization, which is even more challenging.

The first problem is that we may need a licensed or pilot evaluation agency to ensure our certain transaction. The real transaction is the data, not other resources. The data exchange or data trading center can register and record the data products as well as verify the authenticity and asset value in a certain data trading.

For example, a company that has spent 5 million yuan in its financial report and bought data may not be admitted by China Securities Regulatory Commission in the future: How can it prove that this data is worth 5 million yuan? Unless there is advance registration and appraisal when you purchase the product and the Big Data Exchange certifies the asset value of this data transaction, then the data asset is likely to be more widely recognized.

Initially, the value of certain data can only be priced by the demander who is willing to pay for it. To a certain extent, it will be a standard price, such as the data price of renting house in a district, city, and county of Lianjia company. For example, given a remotely sensed data with an accuracy of every 10 square kilometers, it can form a standard and gradually become transparent, and then gradually build up a system for pricing evaluation.

It is similar to our original pricing evaluation for intellectual property, which finally is reflected in the financial report to see how much value is generated from the intellectual property. We can price the data by using this method and realize the asset and capitalization of data with a licensed and credible evaluation authority based on the ownership or use rights of the data and trading records.

Data has to go through three progressive stages of resourceization, assetization and capitalization, which is a very valuable and at the same time very difficult and long road. My point is that the road ahead is hard and long, but we will be successful as long as we keep going.

How to Land the Market Circulation of Data Elements

Tradewind Think Tank: How do we price data? What principles need to be followed in this process?

Zhou Tao: There are several ways to price data. One is that the data has transaction records, so it is relatively simple to fix the price. It is the data I sold to you that has already been sold to others before. And you are willing to pay one million yuan to buy this data, at least the third-party evaluation authority can evaluate that the data you bought is worthwhile. If you spend one million yuan buying data and write it into your financial report, there is no problem.

It is also easier for standardized data products. Because the industry has its own standards, we have a set of pricing methods such as big data from remote sensing.

The most difficult thing is actually how to price a set of data. In the past, companies could make a lot of money by making some products, while now companies have to price the data generated in the production process like R&D inputs or other assets, indicating that the data are assets.

In this aspect, it can refer to the evaluation methods of intellectual property in enterprise transformation and confirm the related project with direct profits according to the enterprise's financial report, which still requires a systematic project, just as it took us many years to pilot it before the market really recognized the intellectual property evaluation.

Tradewind Think Tank: Our country has listed data as one of the seven major market factors equal to land, labor, and capital. Compared with other elements, what are the particularities of the data elements? What are the fundamental conditions for the market circulation of data elements?

Zhou Tao: First of all, data has some characteristics of essential productive factors, such as its scarcity. But the scarcity of data is not the same as that of land for the certain amount of land will be less after being used, and the quantity of data will be increased after being used in a sense, which can be replicated more back and forth.

But there is still lack of data. For example, when we want to use remotely sensed data or e-commerce data, although the data can be copied endlessly, we still have to spend money buying it if we don't have one. Therefore, the quantity of data will be increased after being used, but scarce. Scarcity is a key factor of production, without scarcity, it cannot reflect its important value position in an economic circulation.

Another feature of data is that it is not ownership that is often transferred or traded when we transfer or trade data. Other elements such as land, often transfer their ownerships in essence. Intellectual property can transfer ownerships and the rights of use, transferring ownership for investment in most cases, or transferring the rights of use for authorized redevelopment. In the vast majority of cases, the data transfer the rights to use, not the ownerships.

For example, how do people judge whether the data sold to us will resell or not? There is no good way for the moment to judge it. We can put the data in a medium, so that we can only use it in the medium, such as a CD or a USB flash drive, we can only use that USB flash drive in the computer. But, it is not a good deal as it is very hard to use.

To avoid repeated purchases of data, we need to have a trading center to keep the rules of data transactions. Since the data transaction center can not only record transaction behaviors, but also ensure that the resale cannot be registered and is no longer valid or cannot be traded. Therefore, we also need to establish various mechanisms to ensure that we cannot give it to others while giving us the right to use the data.

There is no way to trade the ownership of data because the ownership is unclear. It is difficult to achieve data validation in the short term with little significance. For instance, A company collects a lot of our personal data and it is clear that the ownership of the data belongs to consumers or A company. Or if the government commissions a company to provide a free service and we agree to contribute data, it's hard to say whether that data is the company's, the government's, or ours.

If we want to make it clear in the short term, it might hit the big data industry and it has no practical meaning and value for individuals to obtain the ownership of the data. Some people may wonder if individuals after submitting data will earn a small amount of money, which I think may not be sustainable, because it may not have many benefits to individuals, nor to obviously boost the industry.

In this sense, as a productive factor, data will keep in a state of unclear ownership during the transaction for a long period of time. Therefore, what we need to guarantee is not the transaction, but the act of transaction, which guarantees its value and uniqueness.

In general, it is really a new thing and a new factor of production, so it requires us to think deeply about law, even ethics.

Tradewind Think Tank: The data trading center is the carrier of market circulation of data elements. Recently, data trading centers in some places have begun to use new technologies such as privacy computing to make data available but invisible. What impact will this have?

Zhou Tao: Data ownership and the rights of use can be separated in some cases, and unclear in some conditions, and even difficult to separate in some circumstances. Then, to secure multi-party computation, federated learning, and private computation are actually solving some problems to some extent.

For example, several banks provide the service of data sharing. They may pay for each other, or be free of charge, and then through secure multi-party computation or federated learning, each obtains the required calculation results, at the same time preventing other parties from knowing their customers' data.

However, in most cases, the data provider is single who does not want the users to take the data, which is unnecessary to take the way of federated learning or secure multi-party computation, but requires to use data sandboxes and privacy computation. You run the program in my data sandbox instead of taking my data, only getting the calculation results.

These new technologies can only settle parts of the problems. There are still many situations where I have to use your data to calculate. Because I may not be able to figure out what to do with data. I will do research based on your data, but I don't want you to know what I've calculated, which probably is the trade secret, so I must take your data.

But after I take your data, it is the most difficult problem that you do not expect me to sell it again, which cannot be tackled for no particularly good solution now. On the whole, we rely on current technology to gradually clear obstacles for data transactions. For example, API interface queries and calls are currently easy to use at least.

How Big Data Empowers Social Governance

Tradewind Think Tank: What are the main problems solved by the data management and supervision recently discussed in the industry and what significance and value does it have for the development of the entire big data industry?

Zhou Tao: Data management and supervision is actually a kind of professional supervision service. At present, in the construction of a large number of information systems, the owners require these information systems to generate some data, but no professional organization are available to judge that whether the system can produce quality data and whether the data meets the corresponding standards. It is actually a very professional system, including the judgement of the data logic, value consistency, and exceptional detection.

When we used to make an information system, we often only controlled and tested the parameters of information. At the same time, we would estimate whether the system satisfies the clients' business processes and requirements. Usually, the system could meet the business requirements, but it does not meet the data requirements.

For example, a government department has built a service system for common people to reflect the living conditions of the citizens. They can take photos and upload them, and the government can handle it. But if we want to collect all the photos that involve unevenness or cracks on the ground, we may not be able to do it. Many systems can only right-click to save one by one, which means that if there are 100,000 photos, you have to right-click 100,000 times. From the perspective of data supervision, such a system design cannot smoothly export data according to certain requirements.

Another common situation is that although we can describe the data of photos to form high-quality data annotations, the description and the photos are separated in the database, therefore, we cannot link them to form a valid data set.

Of course, there are many reasons leading to the poor data quality. For example, the system design might intend to make people look good, no matter how the photo is taken, it is forcibly displayed in a frame. If you understand data management and supervision, you should establish two systems of photo data, one to adapt to the display frame, and the other to keep the original high-definition photos from distortion.

If we want to ensure the formation of high-value data, it is essential to form a third set of indicators that are independent of information and business indicators, and then form a professional company and service team to ensure that the third set of indicators is well completed, which is to be achieved for data management and supervision.

Tradewind Think Tank: In terms of COVID-19 epidemic prevention and control, what roles does big data play?

Zhou Tao: Overall, the application big data in epidemic prevention and control can be divided into three levels: The first level is to meet the precision of emergency management. For instance, when the epidemic broke out, we quickly had health codes all over the China and shared information with the three major communication operators to know how long a person has stayed in a certain place, as well as the general track and location. At the beginning of the epidemic, big data accurately found all people from Hubei, which was beneficial to epidemic prevention and control according to different situations.

Later, in large cities with relatively high levels of informatization such as Shanghai, Hangzhou, and Chengdu, although the COVID-19 epidemic has sporadic outbreak, cities and streets were closed. There was a sporadic outbreak in Pixian County, Chengdu. All close contacts were immediately found and completed nucleic acid test, which did not disturb the order of normal work and life of the people. But as soon as the epidemic broke out in Ruili, Yunnan, the entire city was locked down, and everyone in the city had to complete nucleic acid test, because the latter one has not collected sufficient information. The second level is to grasp the general track and adopt technical methods that enable us to find the clse contacts, in addition to the basic information.

The third level is to do a social simulation of the entire prevention and control of infectious diseases. For example, when we collaborated with Italy, where the epidemic was more serious, we divided the city into grids of 0.25×0.25 (latitude × longitude) to count how many companies, schools, and families were in the grids. For each person in the grids, based on real statistics, we carried out stochastic simulation based on the method of Markov Chain Monte Carlo (MCMC).

Then, we could judge the possible impact caused by each act and each policy. For example, we had to find out the reason that the closure of Italian schools did not really ease the epidemic. We found out that the method they used was to close the entire class if one student was infected, and closed the school if three students were infected, but the virus had already actually spread. We later recommended that antibody testing be implemented on a certain time cycle, and based on the antibody testing to determine how the school needed to be closed, which ultimately shortened the closure time of the school, and the epidemic situation was better controlled.

What's more, we are studying epidemic prevention and control measures in public areas. Based on each place, such as a supermarket, we will study how many close contacts it may have, and then decide whether to close the place. Here we have to look at three indicators: The first is the impact on the citizens, the second is the extent to which how many close contacts can be reduced, and the third is the economic impact.

We build a digital twin of the real society through large-scale simulation based on real data, and then we can test various policies to simulate their impacts on epidemic prevention and control so as to precisely choose the best policy.

Tradewind Think Tank: How does big data play a role in modernizing social governance and building a new grassroots social governance system?

Zhou Tao: The key to the use of big data in social governance is to use big data to really figure out how many resources there are, where these resources are, and how efficiently they are being used. As long as we can figure out these problems, we can really find the crux of the problem, instead of blindly increasing some unnecessary investment.

For example, based on the data of epidemic prevention and control, we may begin to gradually know how many people live in a city, how many people work in the office buildings, and whether a large number of newly built office buildings are in use, and its accurate utilization rate.

Another example is the construction of public toilets in a village. We need to find out the numbers of public toilets in a village, how many people use it, and how often it is used. We can't make an impulsive decision.

Therefore, if we want to make the best use of big data, the first step is not to use intelligent means, but to really use these data to figure out what resources are available at the grassroots level, which are to be rebuilt and reinvested, and which just need to be revitalized. For example, are there enough libraries, youth centers and universities for the elderly at the grassroots level? These are what the data can provide help.

How to Deal with the Data Explosion Brought About by the Internet of Everything

Tradewind Think Tank: How should we respond in terms of data governance and security when facing with the future data explosion brought about by the Internet of Everything?

Zhou Tao: Compared with the traditional Internet, the most crucial difference between the Internet of Everything is not the source of data, but whether the bulk of data comes from a controllable scenario or is merged from multiple scenarios.

For example, a Chinese corporate Alibaba has a lot of Internet data, but they all come from Ali. These data can be directly served to everyone through the Taobao and Tmall platform as long as we manage this company well. These services can give the enterprise financial returns, but also make consumers satisfied.

But the Industrial Internet is completely different because data comes from different factories, and the management participators between these factories are different. If we were to use one platform to collect or manage this data, we would face extremely high complexity and different data standards, so it would be less likely to generate value.

Every factory has distinctive features, environments and products. If one factory provides service from one factory to another, its essence is to do the project and without general applicability. If all these data are taken and applied for providing an inclusive product or platform, it will be very difficult, which might have to be negotiated from one factory to another, and then becomes a project again.

In fact, this is a question of how to manage in the case of decentralized and large amount of data. For scattered factories, it is difficult to manage from the process. For the moment, the management is based on results. The party of management provides safety standards and technical services, which can protect the factory. If the factory's problems cause damage to others such as data leakage, the factory must be punished for the consequence. In the way, it is a bit similar to managing a large number of pollutant companies in environmental protection, providing services to standards, and finally imposing penalties or rewards based on the results of emissions.

In addition, the situations such as unmanned driving, pacemakers, or robotic arms must be incorporated in specific scenarios for management. Because these scenarios have special safety hazards and requirements, once a problem occurs, it may cause serious casualties.

Tradewind Think Tank: In the era of big data, how can we enjoy the convenience brought by big data without being manipulated by it?

Zhou Tao: Why are people being manipulated? Because when being manipulated, we will feel very happy and get a sense of satisfaction at the spiritual level, thus put ourselves in a kind of information cocoon. For example, in foreign social networks, we will find that in the U.S. 90% of the accounts followed by Republicans are Republican members, and only about 10% are cross-party followers. When people choose access channels of information, they will actively choose the same or similar people, as on such a social network, people will feel happy, and everyone will respond to anything they say. Therefore, we can't completely blame big data for all problems as we are voluntarily manipulated in many cases.

In the second case, the merchants manipulate people for profits. This situation is not always voluntary. For instance, we actually want to discover some new things on some news websites or e-commerce platforms, but sellers repeatedly deliver contents that we have a strong interest in, or things that are most likely to be purchased. In this case, information technology becomes a convex lens, making our vision narrower than before, which is called information cocoon.

How to break the information cocoon? We are currently doing

some researches to make our horizons more open through a better information navigation technology that increases in diversity without affecting the click-through rate. However, at what level these new technologies can be recognized, and whether these new technologies will become the mainstream of the industry, which is still hard to say.

Tradewind Think Tank: People in the industry often mention "Data makes decisions," on the contrary, it is the problem of information cocoons. Do we have a reference to make decisions that need to be done by ourselves? And which decisions can be made by big data and artificial intelligence?

Zhou Tao: I think we have to make all decisions by ourselves. The behavior of a machine is not called a decision, but a response that according to a fixed procedure. What we hope is not to allow the data to make decisions, but to make the data speak and tell us the most important information, and it is still us who make the decisions.

When Edison invented the electric light bulb and founded General Electric, the electrification was far from widespread. General Electric's early investors distributed questionnaires and asked the public: If there is something like electricity, what use would you think? The result was that everyone felt that it was useless because many people had no concept of electricity. How could they know what it was? Even Edison felt that the only use of electricity was to make electric bulb light.

How to make decisions at this time? Should we invest in

General Electric, or promote electrification businesses? In this case, it is impossible to only rely on the data from surveys.

Data can help us make general decisions, but not creative, disruptive and major decisions, let alone make decisions that require self-knowledge. For instance, big data can't help us choose a wife. Although it seems there is a set of methods to make the choice available, the reality is that whether two people are in love after a relationship cannot be judged by the data.

I have cited an example for many times: The World Cup in 2006, the German goalkeeper Lehmann learned about the opponent's shooting angle preference for penalties through statistics and saved multiple penalties from the opponent. However, it assumes that the goalkeeper knows the opponent's shooting angle preference by statistics, while based on experience, he has judged that the opponent must do the opposite. Under this circumstance, it would be unreasonable if the defense should be carried out according to the results of the data statistics.

Therefore, what data gives us is always objective judgments that is not the same as decision-making. It is mankind that makes the final decision.

Wu Xiaoru:
How Intelligence Facilitates Life

How can the system innovation of artificial intelligence solve the major propositions of social development? What are the cognitive misconceptions and practical shortcomings in the application and development of artificial intelligence? In the process of AI application, what kind of goal should we take as the guide? What kind of principles should be followed? How can artificial intelligence companies avoid the possible negative effects of technology and realize the true sense of technology for good?

What Is the Next Step of Artificial Intelligence Innovation?

Tradewind Think Tank: iFLYTEK is currently committed to "using artificial intelligence system innovation to solve major problems of social development." How do we understand "artificial intelligence system innovation?"

Wu Xiaoru: This is what we have been thinking about: How to solve major problems of social development through AI systemic innovation. We consider it from two dimensions. One is the innovation of single point technology. For example, in speech recognition technology, we continue to break through the difficulties of speech technology and continuously improve the accuracy of speech recognition. At the same time, we have developed a series of derivative products with speech technology as the core such as

"Xunfei Hear" and "Recorder." Here, we mainly consider problems from a technical point of view. Once we make breakthroughs in important technologies, some problems can be solved.

However, solving a single technical problem is not the full picture of innovation. Technological innovation has presented a new trend of generalization, serialization and structure. The second dimension starts from the problem itself. We aim to find out which technologies we can integrate to form a solution.

When we go deep into some important industries such as education, medical and transportation, we must have in-depth exchanges with customers in these industries. We need to think how to solve problems in these industries through technology, so as to reduce costs and increase efficiency.

Take the education industry as an example. The education industry does not care how innovative your technology and speech recognition are. Their actual pain points are the heavy burden on teachers and students, and the uneven distribution of urban and rural educational resources. These are major problems in social development. Therefore, we have to analyze the factors that cause these problems and look at the technology with the intention to solve the problems.

For example, how to reduce the burden, like heavy lesson preparation and homework correction, on teachers? This could by no means be solved by a voice recognition or image recognition technology. There may be a series of technical application problems behind. Some of them can be solved by existing technologies, and some cannot, which requires us to carry out technical research.

Starting from these two dimensions, we proposed the systemic innovation of artificial intelligence. The purpose is to understand what problems we have to solve in the end, what are the reasons for these problems, and what kind of technology could deal with them, so as to plan and solve them step by step.

Influenced by the change from single technology to system innovation, our requirements for talents have also changed. In the past, we mainly brought in technical personnel. Now we need inter-disciplinary talents who can deeply understand the industry, and solve problems by combining industrial problems with AI technologies. It is a very important challenge for the further application and development of AI technology.

Tradewind Think Tank: Looking back on the applications and development of artificial intelligence over the years, what are our cognitive misunderstandings and shortcomings in practice, and how can we make up for and improve it?

Wu Xiaoru: Current artificial intelligence application is based on deep learning. Technology embodies theory and then extends to solve social problems. But artificial intelligence is not a panacea. We should be rational and recognize that there is still a long way for artificial intelligence to reach the so-called "intelligence," although it is already very powerful in specific scenarios.

In 2017, China set up the National Key Laboratory of Cognitive Intelligence in iFLYTEK. Cognitive intelligence is different from perceptual intelligence. Speech recognition and image recognition are perceptual intelligence. Cognitive intelligence is more advanced, and it is the artificial intelligence that understands what speech, text and images are trying to say. In terms of the current theoretical foundation and technology , it is basically impossible for us to build a powerful, general AI model that achieves cognitive intelligence and solves all problems, unless there is a major theoretical breakthrough in artificial intelligence.

Of course, the existing artificial intelligence is data-driven intelligence. When applied in some specific scenarios, it can solve some practical problems in a simplified manner. When it comes to the diagnosis of a specific disease, medical experts can gather to collect a large amount of data for this specific disease, and to form the relevant pathological diagnosis and treatment experience into a knowledge map that can be processed by artificial intelligence. In this way, such specific disease could be treated. Therefore, even without theoretical innovation, we can combine a large amount of data, industry knowledge and AI algorithms in a relatively small field to form a highly targeted solution.

Currently, artificial intelligence is in such a preliminary stage of combining specific scenarios to solve specific problems. The same is true for iFLYTEK's speech recognition technology. Our goal is to recognize all dialects and languages, and to combine specific application scenarios and industry fields for more accurate recognition.

In fact, as early as 10 years ago, iFLYTEK has already been studying deep learning algorithms. We are still very much looking forward to a major breakthrough in theory, even if there is no breakthrough in building a framework, and hope that the existing theory can continue to move forward. Therefore, iFLYTEK has now established laboratories jointly with some universities. Especially in recent years, we have conducted in-depth cooperation with the mathematics departments of universities, and some returnee mathematicians, hoping to further develop mathematics and science. Once there is a clear direction for theoretical breakthroughs, artificial intelligence will progress faster.

However, it is a long process to achieve theoretical breakthrough. It is more important to combine technologies of artificial intelligence with application scenarios and to turn industrial knowledge into certain technology to solve problems in specific scenarios. We must be rational to see that current artificial intelligence is far from being able to solve some problems or even sometimes surpass human intelligence, it is still able to solve some problems.

What Fields Can Artificial Intelligence Explore and Apply to?

Tradewind Think Tank: In terms of basic research, what cutting-edge explorations has iFLYTEK made in artificial intelligence?

Wu Xiaoru: People may wonder what is the difference between the artificial intelligence technology developed by iFLYTEK and that by other companies? In fact, the artificial intelligence technology framework is also layered. The first layer is to develop products with existing artificial intelligence technology and position itself as an artificial intelligence technology company. The second layer is to develop some speech or image recognition products through some open framework source of artificial intelligence technology, but they can only do within the framework. The third layer is the entire deep learning algorithm for speech or image recognition, which is completely developed by itself. In this case, some adjustments could be made based on the theoretical framework, such as making adjustments on the network structure that simulates the human brain. The fourth layer is combined with some basic researches on mathematics and physics. In fact, to a certain extent, deep learning is to extract effective information layer by layer from abstract, disorderly information, so as to improve the information extraction ability of this model or network. Researches on the basis of mathematics and physics aim to highlight some of the original representations of certain algorithms of deep learning and to solve the interpretability problem of deep learning and other abstract problems.

iFLYTEK is working on the fourth layer. We are collaborating with research teams in the field of mathematics and computer. We cooperate with Beijing Normal University to study brain-like cognition so as to make breakthroughs with the cross-collision of different disciplines. We can get good rankings in some important international competitions every year because we have original technologies and integrate everyone's basic research strength together. If using an open-source framework, we can only achieve 90 points at most while others reach 100.

In terms of frontier research, many influential projects are being implemented at the national level. We participated in a project of the Ministry of Science and Technology named "Questions Answering of Humanoid Machine," in which the machine takes the college entrance examination to see how many points it can take each year. This can reflect the advance of the artificial intelligence algorithms. For example, how to turn knowledge in liberal arts and science into information that computers can understand. The entire logical chain of artificial intelligence includes the basic description, understanding, analysis, reasoning and decision-making of information. This chain requires a variety of research teams to join in, including mathematics, computer science, linguistics, and pedagogy. Everyone integrates together and advances the research step by step. In this process, there are breakthroughs in computer engineering, in technical theory, in ontology description methods for professional knowledge such as text, history, and geography, as well as breakthroughs in network structure and mathematical modeling. Frankly speaking, these breakthroughs are not as fast as the breakthroughs of computer engineering technology, but we can still see the progress in the underlying research, such as medical pathological diagnosis and correcting the composition for teachers.

This round of artificial intelligence development triggered by deep learning theory originated from the West. In terms of basic research, there is still a gap between China and the West. However, in recent years, China has invested more and more in basic research, and we are also looking forward to faster progress in large-scale basic research, and in large-scale mathematics, physics, and neurology research.

Tradewind Think Tank: In applying artificial intelligence to all aspects of people's lives, what goals should we follow? What principles should be observed?

Wu Xiaoru: This should be viewed from two perspectives. The first perspective is whether the use of this technology creates value such as whether teaching and learning efficiency has been improved in education, and whether it has improved the quality of teaching. No one will accept valueless technology. The second perspective is whether this technology is in line with the long-term needs and policy requirements of the laws of social development. For example, the country's governance in the field of education and training is because education and training institutions only solve some short-term problems. If children are not good at math, they will immediately intensify exercises and training in mathematics. However, after this, children's mathematics ability may not be improved. Therefore, the application of technology must conform to the long-term development law of the country and society, and cannot violate the long-term law and make short-term quick money.

We must consider both perspectives. On the one hand, the efficiency improvement brought by technology should be visible and tangible to users. We emphasize the use of statistical data. For example, how much teaching efficiency we have brought to teachers and students, and how much inefficient or even ineffective time has been reduced.

For example, now the children's homework is all same for thousand people, and the children with a score of 95 and those with a score of 55 are doing the same homework, which is inherently unreasonable. Previously, the teacher had no choice. There were fifty or sixty students in a class, and it was impossible to assign a separate homework for each student. But now, technical means can help teachers give each student a portrait and provide personalized homework. For example, the student with the best grades only needs to do 5 questions out of 50 questions, and the teacher assigns him 5 more difficult questions. To this extent, we can see through statistical data how much time such a function can save for teachers and students, and whether teachers and students are willing to use it. Based on data feedback, we will further optimize the algorithm function and user experience to make it more and more in line with the actual needs of teachers and students. In this way, we can form a virtuous circle between scenes, data and algorithms.

On the other hand, technology must be in harmony with the laws of social development and national policies. For example, education should take focus off score results, and students cannot be indoctrinated for short-term improvement of scores. In fact, from a long-term perspective, the improvement of students' academic performance must be the results of improvement of learning ability. Learning ability consists of language expression ability, logical operation ability and reasoning ability. In this regard, we have conducted in-depth cooperation with Beijing Normal University and East China Normal University to explore the application of AI technology to assist teaching. The research has tested and improved students' learning ability in a short-time and is in line with the long-term law of educational development.

iFLYTEK not only wants to enable machines to listen, speak, understand and think, but also hopes artificial intelligence to build a better world. This is not a slogan. For example, voice technology can be used for telephone robots, which can contribute to epidemic prevention and control, but can also be used to make harassing calls. So, we must adopt some technical methods to filter or shield to prevent such things that are detrimental to society.

What Improvements Does Intelligence Bring to all Walks of Life?

Tradewind Think Tank: How does artificial intelligence promote the distribution of educational resources, teach students in accordance with their aptitude, and evaluate quality education?

Wu Xiaoru: This is a good question. The issue of educational fairness is a concern for everyone. There are various reasons for the uneven distribution of educational resources, such as school district housing. But technology can only solve part of the problems.

The first focus is on how to use technology to empower the teacher's ability. A teacher's teaching is oriented to two or three classes, no matter how strong his ability is, he can only teach two or three classes well. With the popularization of online education, teachers can teach more students and no longer be limited to the physical space. How does it work? Teachers can not only transmit knowledge to students through the network, but also can have a two-way connection with students through artificial intelligence. First of all, artificial intelligence equips each student with a learning-assisting robot, which can judge their learning status for the teacher's reference. Secondly, after the teacher completes the teaching, he can send test questions to one thousand or even ten thousand students. Artificial intelligence can assist the teacher in analyzing the test results, thereby optimizing the teaching content and adjusting the teaching progress. In addition, it can set up a dual-teacher classroom. One lecture's teaching is open to multiple remote classrooms through the Internet. And another teaches on site at the remote classroom, which can interact with the lecturer, and pay attention to the students on the spot, putting students in a better study condition. Especially when it comes to language learning, English teachers are not necessarily available in many rural areas. We can use AI technology for speech synthesis to achieve language teaching and evaluation similar to those of real English teachers. We can also use AI technology to grade essays so as to assist rural teachers in setting up relevant courses. These intelligent auxiliary methods can, to a certain extent, allow more schools, especially teachers and students in rural areas, to obtain high-quality educational resources.

The second focus is on how to carry out individualized teaching, we first digitalized the teaching process of teachers and students, such as homework, test results, and teacher-student interaction. Some schools also let students wear bracelets to analyze physical fitness and exercise data. Students' portraits could be continuously provided through intelligent analysis of accumulated data. As more data is collected and the longer it is conducted, the clearer the portrait becomes. Therefore artificial intelligence can provide students with more personalized learning plans. The customized learning plans reduce the inefficient and ineffective learning time for students. The time saved can be used for the cultivation of comprehensive and innovation ability. If we still follow the education model of 20 years ago, the next generation of children may not be able to adapt to the future society. Therefore, we must enable students to learn more accurately in current academic system, and provide innovative learning methods. For example, we are cooperating with schools to carry out innovative education in artificial intelligence and language.

The third focus is on how to build a more comprehensive evaluation system. Although there is no evaluation method that can replace the college entrance examination, China is encouraging to take focus off the score results. A new evaluation system for schools, teachers, and children is needed. For schools and teachers, the number of students admitted by famous key university should not be the only measurement. For students, in addition to the college entrance examination, there are other channels for selecting talents. So, how to change the situation that a student's life is decided by the college entrance examination? It is necessary to form a more all-round evaluation mechanism for students in daily learning performance. This includes two aspects. The first is an efficient and accurate evaluation method, which means big data artificial intelligence must be consistent with the current education system.

The second is a more open and transparent evaluation process. The process should be more socially credible, like using unchangeable blockchain to form an advanced evaluation system with technology as the key support, and we are already making relevant explorations and attempts. When the credibility of this evaluation system becomes higher and more accurate, it may change the current system of selecting talents for college entrance examinations.

Tradewind Think Tank: How does artificial intelligence promote the allocation of medical resources, improve the efficiency of diagnosis and treatment, and help prevent and control the epidemic?

Wu Xiaoru: In 2017, a smart medical robot of iFLYTEK took part in the National Qualification Examination for Medical Practitioners, and outperformed 96% candidates, which made us more confident. Therefore, a robot could deeply understand the ontological knowledge of medicine.

Of course, there are still many problems in the medical field that need to be resolved. The first is the hierarchical medical system. Patients all rush to grade A tertiary hospital, resulting in an extreme situation that such hospitals are overcrowded while the general hospitals have few patients. We hope that we can strengthen the grassroots hospitals through technical means to assist grassroots doctors to make more accurate and effective diagnosis and treatment. In this way, when ordinary people encounter some minor or common diseases, there is no need to go to the grade A tertiary hospital, and they can get better medical services in community hospitals. Doctors at the grade A tertiary hospital only give three minutes of explanation, while doctors at community hospitals will give a detailed explanation of ten to twenty minutes. We call this technology smart auxiliary diagnosis. At present, we have achieved more than 300,000 auxiliary diagnosis and treatments per day, which has improved the diagnosis and treatment capabilities of primary hospitals to a certain extent, and enabled grassroots hospitals to solve minor and common diseases.

The second is daily consultation. Let's take the diagnosis and treatment of chronic diseases as an example. There are a large number of chronic diseases patients in China. They need drugs and other treatments in their daily lives as well as the care of doctors. But obviously our doctors cannot deal with so many patients. Therefore, we cooperated with the hospital to develop an intelligent application system for chronic disease management and medical care. In daily life, patients can seek medical advice through APP or phone. How is the body's current reaction? Will you be allergic after taking the medicine? How is the recovery? How should diet and medication be adjusted? We model common questions that patients ask doctors, and then let the machine automatically answer these questions in an intelligent way. In the past, there might be 100 questions waiting for a doctor. Now the machine can help answer 95 common questions.

The third is some issues related to the national prevention and control of infectious diseases. In the original prevention and control of infectious diseases, the disease was reported level by level when the primary hospitals discovered the epidemic, and the Center for Disease Control would analyze it. Now we have set up mechanisms for infectious disease prevention and control and early warning in many communities. If an infectious disease is found in several communities at the same time, the early warning mechanism will be triggered at these communities, and the CDC can get the early warning report immediately. At the same time, based on the population database of these communities, our automatic epidemic detection system will be used to call everyone in the communities. In this way, the speed of early warning of infectious diseases and the efficiency of investigation can be significantly improved.

Tradewind Think Tank: Automobile industry is one of the most in-depth scenarios of artificial intelligence applications. How does iFLYTEK empower automobiles?

Wu Xiaoru: Artificial intelligence can most stimulate imagination, while automobiles require comprehensive technology application. Therefore, the combination of artificial intelligence and automobiles can create unlimited imagination.

In the past few years, iFLYTEK has gone deeper and deeper in automobile application. We have cooperated with many automobile manufacturers such as SAIC, Geely, and Chang'an, and our voice interaction system has been applied to more than 23 million vehicles of more than 1,000 models. Nowadays, more and more cars are equipped with voice interaction function modules. According to our statistics, every newly manufactured car is equipped with voice interaction function. Among them, seven out of ten vehicles use the iFLYTE technologies.

We can see that more and more artificial intelligence technologies of autonomous driving, voice recognition, image recognition, and cognitive intelligence, are applied to cars, which will make future cars smarter and more considerate.

How could a car be smart and considerate? For many cars nowadays, users can send out a variety of voice commands by pressing a button. Although this application has embodied the artificial intelligence, it is not considerate enough. In the future, without pressing a button, the car can determine at any time whether the current driver is talking to the car or the passenger next to him. When driving, the driver can regard the car as his friend and enjoy a variety of services provided by the car. In the future, the car will be smarter, more considerate and more accurate to accept, and carry out various commands.

At the same time, the car of the future will not only be able to react to commands smartly, but also provide personalized service. In the future, automobile will realize a multimedia, multi-channel, multi-style human-computer interaction scene. By installing a camera in the car, the car can accurately provide a variety of services whether it is for the father who is sitting in the front row, or for the daughter who is sitting in the back row. Unlike mobile phones, cars are durable consumer goods. Mobile phones may be replaced once a year or two, but cars will be used for many years. Therefore, in the application process, the car can constantly understand the preferences of family members and everyone's preferences so as to provide each person with more and more personalized services.

Car is also a stylish and comfortable moving space. Through the integration of AI technology and other technological applications, cars can automatically reduce noise, and voice in calls will be clearer during high-speed driving. In the past, it required long-time tone tuning experiments and the deep participation of experienced experts to get great sound effects, but now it only requires these experts and AI technology to work together to efficiently tune the sound effects to the best. A multimillion-dollar car audio effects can be provided by a car only worth more than 100,000 yuan.

Car is a safe and reliable moving space. For automobile manufacturers, it is necessary not only to develop and build cars, but also to provide users with safer and more reliable services during using. Rolls-Royce remotely monitors the operation of aircraft engines through the network and algorithms. Likewise, automakers can also establish connections between cars and automakers, and provide full life cycle services from the purchase to the sales and to the use. The desensitized static and dynamic data of cars enable car manufacturer to know the running status of each car in real time, providing timely fault risk warning through algorithms and give emergency treatment plans in time, so as to ensure the safety and reliability of each car.

From the perspective of artificial intelligence, future automobile is an AI capability platform that integrates visual, auditory, and dialogue systems. We can desensitize the data of each car and each user on the car through the network, and store it in the user data platform. In this way, the integration of artificial intelligence, big data, and cloud computing advanced networks can further make the moving space more intelligent, give the car a smarter brain, and provide each user with tailor-made services and a better travel experience.

How Does Artificial Intelligence Shoulder Social Responsibility?

Tradewind Think Tank: How can artificial intelligence better serve the elderly and help them better embrace intelligence?

Wu Xiaoru: We have been paying attention to this issue a long time ago. However, developing some smart applications for the elderly are still not easy. In this regard, we have considered and designed products with three dimensions. Some have already been finished, and some are still in further development.

The first dimension is how to make existing intelligent services senior-friendly. Young people can easily use various websites and apps, but it is a challenge for the elderly. The government also attaches great importance to senior-friendly applications, and reforms government services and public services to adapt to the aging population, making it easier for the elderly to apply for certificates and subsidies. For example, voice is the easiest and most natural way of communication between people. The elderly can easily use voice to consult various services and ask various questions. We cooperate with some local governments to intelligently transform existing government hotlines and apps to make it senior-friendly to integrate into digital lives.

The second dimension is to provide health and safety monitoing and assistance for the elderly. In many communities now, the elderly people are alone at home without children around. It is impossible for community personnel to knock on the door every day to care about them. But it is possible that the elderly people get sick or even pass away that no one could know. Therefore, we made some intelligent perception systems, such as installing a camera outside the apartment of the elderly to connect the utilities with the management system. If the system finds that the apartment has not used any utility for several days and the elderly has not been out for several days, the system will automatically call the children and community workers to tell them to check the elderly.

The third dimension is to alleviate the loneliness of the elderly to a certain extent. Elderly people living alone cannot communicate with their children every day. To solve this problem, we are exploring some senior-friendly services and products to provide emotional communication. This involves humanity and ethical problems, and we are still exploring. It is also difficult to continually provide some services and products that the elderly will be willing to use for a long time. Now senior-friendly technology has become a social concern. We believe that in the elderly society and the "silver economy," artificial intelligence and robots can do more. There are already some hotels where robot will deliver some toothbrushes or fruits for customers. Applications like these can be used in the life services of the elderly community. Therefore, we open the AI technology to society and help more partners and technical teams develop a variety of intelligent senior-friendly applications for the aging population through the open platform of iFLYTEK.

Tradewind Think Tank: As an AI company, how to avoid the negative impacts that technology may bring, and ensure the use of science and technology for the greater good?

Wu Xiaoru: Actually, this is a question of corporate attitude. One attitude is that I will do everything that can make money as long as it does not violate the law; the other is that we hope to use our technology on the positive side of the society, but if there are some negative effects, even if the law does not prohibit it, we will absolutely not do. This is the basic attitude of an enterprise.

We at iFLYTEK have established the corporate vision of "Let artificial intelligence build a better world," and have promoted and implemented this vision continually from technology research and development to product services. We would never do things that are not good for society, even if it is very profitable.

With this basic attitude, we have established a set of codes of conducts and methods that must be followed internally. For example, in terms of data privacy, we strictly require that the collection, storage and use of data are absolutely not associated with personal identity, curbing the privacy leaks from the source. For another example, when we are conducting data modeling and building algorithms in the education field, we will hire educational experts and scholars to inform us what should we pay attention to when we take personal portraits of students and provide personalized learning programs, and which aspects will increase students' mental anxiety and study burden.

These all reflect the basic attitudes of the company. We do not just hang these slogans on the wall. From data security and data privacy, to the codes of conducts for each employee, to the final products and services, the corporate culture and basic attitudes run through in every part.

Xiao Feng:

How Blockchain Builds a New Type of Production Relationship

How to view the new production relationship represented by blockchain? How to understand the Internet of Value represented by blockchain? What fields can the distributed technology in blockchain be applied to? How will the Token Economic Model affect our production and life? How can blockchain play a role in data trust and security? What will be the impact and significance of the sovereign digital currency issued by China?

What Is New in the New Production Relations?

Tradewind Think Tank: In the era of intelligence, there is a consensus that artificial intelligence is the productive force, big data is the means of production, and blockchain represents a new type of production relationship. In your opinion, what is new about this new type of production relationship?

Xiao Feng: Blockchain can be viewed from two perspectives. From a purely technical perspective, it is a productive force itself. From the perspective of economics, blockchain technology will indeed play a very new and significant role in reconstructing production relations.

Blockchain cleverly introduces a lot of game theory, especially involving a game theory-based mechanism design, such as distributed ledgers and smart contracts, which is a very important topic in economics.

Therefore, we can say that the blockchain has changed the production relationship, which is a very important change. In fact, it is evident from crypto currencies. There are thousands of people in this world who can design electronic money systems like crypto-digital currencies. But if this is not done based on the blockchain, it has no credibility. Because internet-based electronic money systems have always been available. Only blockchain technology can ensure the certainty and uniqueness of electronic currencies. The blockchain network is then able to confirm the ownership of electronic currencies and no longer requires a third party to do so.

How can you say that a string of characters stored in a computer must be valuable? How to make everyone recognize the value of this string? You cannot send currencies to ten thousand people in a time. However, the distributed ledger of blockchain can synchronize the value of this string of characters with 10,000 people at no marginal cost, just like sending e-mails.

It can be said that the blockchain has reorganized many things, such as the circulation of value and the organization. Why do companies exist? Because we need an organization to better gain trust and engage in economic activities.

Mr. Coase, the Nobel Prize laureate in Economics, once pointed out that the company as a form can reduce transaction costs. But if, through a digital technology system like blockchain, it is possible to reduce transaction costs to zero, then the existence of the corporation is not a necessity. Because even without the company, we can still reduce the transaction cost to such a low level that there is no friction in economic exchange. It is from this perspective that blockchain technology brings about a reorganization of production relations.

As we face the tide of digital economy, blockchain technology has brought us a new way of organization, which is perfect for digitalization and digital economy.

Tradewind Think Tank: There is an argument that the Internet in the traditional sense is the Internet of information, while blockchain is the Internet of Value. So, what exactly is this value in?

Xiao Feng: The network has actually gone through three stages. The earliest started with the invention of radio. Human society first constructed a communication network, and telegraphs and telephones followed. Later, from the second half of the last century, the Internet was invented.

What did the Internet bring? It brought the information network. Apart from communication, we can also process data and e-mails. The information network has brought great changes to human society. Whether it is Alibaba, Tencent, Google, or Microsoft, they were all born in the information network era.

We are now moving from communication networks and information networks to the third stage—value networks. Value networks are not just about the value of money, but based on the fact that value networks can help people manage many resources better. In the age of information networks, it is very expensive to manage these resources.

For example, an e-commerce company like Ali sells goods to customers who are totally strangers. To guarantee the transaction, we need a third-party system like Alipay. E-commerce cannot live without a third-party guarantee system, so Taobao emerged in 2003 and Alipay emerged later in 2004.

In the era of value networks, for example, on the blockchain, value exchange and information interaction do not require third-party guarantees. Isn't there a need for a guarantee? Yes, it is required. But with what is the guarantee? We use blockchain for the guarantee, and here it obviously relies on a technical system with mathematical rules. This is the core of the value network.

As we mentioned earlier, all digital activities are frictionless and their marginal cost is zero. The cost of using a zero-friction system for guarantee is obviously much lower than that of a huge third-party network system. As the times move forward little by little, today we refer to the blockchain as the value network. In fact, it can also be called as resource management network.

Tradewind Think Tank: As you said, the Internet of information has given birth to Internet giants such as BAT, so what new business models will the Internet of Value give birth to?

Xiao Feng: When moving from the information Internet to the value Internet, or resource management Internet, the business model is vastly different from the previous information Internet era. There is a good case for us to understand what is Internet business and what is blockchain business.

Coinbase is a digital currency exchange company in the U.S. It was established in 2012 and listed on the NASDAQ in April 2021, with a market value of over $100 billion at one time. What is $100 billion? The total market capitalization of the NYSE and the NASDAQ combined is approximately $100 billion.

Coinbase has 1,400 employees running such a network. But at the same time, there is an identical platform system on the blockchain, Uniswap, which is also engaged in digital currency transactions. It is built entirely on the blockchain, and only has 11 employees. What is its maximum 24-hour trading volume in a day? It is more than $7 billion, which has reached or even exceeded Coinbase's trading volume.

But do you know how many lines of code did these 11 people write for the Uniswap system? 500 lines of code. 11 people with 500 lines of code. That's blockchain business.

Coinbase was established by means of Internet, so it requires 1,400 employees to write tens of millions of lines of code, and a company responsible for information security from start to finish, for system operations, for managing client assets, for brokering client transactions, and for clearing and settling.

While based on the blockchain, 11 people with 500 lines of code in Uniswap did the same thing. Do these things that Coinbase did become unnecessary? Not really.

In terms of the Internet, we need to build a system in all aspects throughout the process, and we have to take all the responsibilities. However, the blockchain provides everything, such as trust. We call the blockchain a trusted machine because before two people make a deal, they must first trust each other. The process of generating trust, the security guarantee, and the role of wallet is done by the blockchain for you, so what you need to do is put your assets in a wallet based on the blockchain. Since the blockchain is inherently a settlement network, settlement and clearing are done by the blockchain. So you just need to write the trading rules clearly in 500 lines of code.

This is a huge difference between Internet business and blockchain business. This difference is not a 10-fold increase in energy efficiency. In the case of Coinbase and Uniswap, this is a 100-fold increase in energy efficiency. So, we can predict how great the innovative capabilities of many blockchain-based applications will be in the future.

The Internet of Everything Needs to Be Decentralized

Tradewind Think Tank: Distributed technology is a big feature of blockchain. It is not only a technology, but also a concept and a model. With the development of the digital economy, what areas can distributed technology be applied to?

Xiao Feng: Distributed business, or distributed economy all originates from a distributed network at the bottom of the blockchain. In contrast, there is a centralized network in the traditional sense. When the network becomes more and more complex, more and more things need to be processed, and more and more areas need to be involved, the centralized network can no longer carry it.

That's why large Internet companies have gradually moved

towards distributed networks and distributed storage when they build their underlying architectures. This is because the network world is becoming more and more complex, and you cannot use a centralized technical architecture to carry such a complex system and such a huge amount of data.

What will happen when 5G and IoT technologies are developed and people and devices, devices and devices, are connected to the Internet? According to authoritative statistics from IBM, there will be 100 billion devices connected worldwide by 2045. Networking is not just about connecting networks, but about collecting data from these 100 billion devices through sensors, which are generated minute by minute and transmitted to each other. Can the centralized network handle it? No, it can't. So it is inevitable to go to a decentralized distributed network.

A very simple example is the smart car, which is very popular now. Now there are two routes, one is to realize the intelligence of the car itself, and the other is the vehicle to everything (V2X) that China is leading. So, when it comes to V2X, do we upload all the data to a data center, let it be processed and then downloaded back? Definitely not. We have to exchange data directly end-to-end as the car passes and immediately make the right decision whether to brake, slow down, turn left or turn right. This is a completely decentralized network.

When the world is so big that 100 billion devices are communicating with each other, it is unlikely that more than 80% of the communication between devices will, and should, be transmitted to a unified data center. Because by the time it finishes the calculations and tells the terminals the results before giving instructions, there could be countless car crashes.

In this case, the distributed network, as we currently understand it, is no longer sufficient to support such a complex system. So it can only be decentralized, let this car and the roadside device exchange data directly, then immediately make a decision, and then drive 10 meters further, and another roadside device tells this car that an old lady is about to cross the road. This car has to make a decision immediately on the spot.

From a network technology perspective, the move from centralized to distributed and then to decentralized is a technical architectural change at the very bottom. This is because as the world becomes more and more digitalized, networked and online, there must be a new underlying technical architecture to match and support it, and blockchain is precisely a tool that can help us to build a decentralized network.

Decentralization does not mean that regulation should be removed, nor does it mean that there is no need for legal constraints. Rather, the world in the age of intelligence needs such a decentralized network to manage, which of course has rules in it. For example, for the V2X system, the state must introduce relevant laws and regulations, including how smart cars deal with unexpected situations on the road.

Think about the future of the IoT, with so many sensors in all devices, on this basis, it will give rise to a distributed economy and distributed business, which will of course be decentralized. Because if the underlying technical architecture and resource management model are decentralized, the economy, business and finance will also become decentralized.

Tradewind Think Tank: Decentralization is a means rather than an end to solve the complexity problem, and we cannot generalize. So, in the actual application process, should decentralization be divided into fields, scenarios and phases?

Xiao Feng: Decentralization cannot solve all problems. Not everything in this world needs to be decentralized. Just like we have experienced the agricultural revolution, the industrial revolution and the information revolution, is there no more agriculture now? There still is.

From the perspective of efficiency, sometimes the more centralized the more efficient and the more decentralized the less efficient.

For example, using the existing blockchain for payments or remittances is an inefficient and expensive thing to do. For a platform to confirm 500,000 transactions per second, it has to use a centralized structure like Alipay.

Because the transaction on Alipay is confirmed by one person, the efficiency is the highest. On the blockchain, hundreds, thousands, or even tens of thousands of people and nodes are required to confirm the entire network, which is obviously inefficient.

There are still many things in this world that require efficiency, but some things also require fairness. Some things may require everyone to witness to reach consensus, but not all things necessarily require a consensus.

For example, the People's Bank of China is designing and testing a central bank digital currency. In fact, the central bank made it very clear that the RMB digital currency does not currently rely on blockchain technology. Why is this still not the same?

The People's Bank of China is legally authorized to issue currency. It is the entire country that does credit endorsement and blockchain is no longer needed for credit endorsement. But in circulation after issuance, the central bank digital currency may also need to use blockchain technology.

How Will the Token Economy Affect Our Production and Life?

Tradewind Think Tank: You mentioned earlier that blockchain is a way to allocate resources. From the perspective of economics, how will the token economic model represented by the blockchain affect our production and life?

Xiao Feng: Good question. In fact, I think the original translation of token into a "certificate" is not appropriate. The development of the blockchain has given birth to the recently popular Non-Fungible Tokens (NTF), which is a very pure certificate.

It is found that blockchain today is very mature as a tool for resource management and value exchange. It includes three categories.

The first is a digital currency, like a crypto-digital currency, issued entirely based on the blockchain, called Coin, which is not the same as a national sovereign digital currency issued by the People's Bank of China, but is an original digital currency with some of the properties of a currency.

The second type is called token, which became purely blockchain-based token after the NFT was separated out. Tokens must first be standardized, that is, the tokens of the same thing cannot be different. It must be a very standard value measurement tool, representing a certain right. Tokens are distributed based on crypto-digital currencies and are more like digital assets than the first category of digital currencies.

The third type is NFT, which is a sheer certificate. It is non-fungible and unique. So, what is NFT used for? To confirm rights, for example, it can be used to clarify the ownership of things. We map a painting to an NFT and whoever has this NFT owns the painting.

In addition to confirming rights, NFT is also used for authorization. For example, in the access control management of an office building, everyone has a pass, which must be non-homogeneous. A visitor can only use it once, while people work in this building can always use it. And people working on different floors can only enter the corresponding floor and cannot go to other floors. This is a pass, or can also be called a token.

Till today, blockchain has developed three perfect tools, digital currency like Coin, digital token like Token and digital certificate like NTF to perfectly help us structure a distributed business or economy based on the blockchain.

So back to the question earlier. What exactly has the blockchain-based token economy changed? Or what kind of dramatic change has it brought?

Let's take two entrepreneurial projects as examples. If you start a business via the Internet, you need to write a business plan to seek venture capital. If venture capital believes in your business plan and team, then they have to talk to you about a core issue, what is the valuation? If it is 10 million RMB, I will first give you an Angel investment of 1 million RMB. And after half a year, if the business model of your startup proves successful, the venture capital will give you a valuation of 30 million RMB and a new round of investment.

But the entrepreneurial logic on the blockchain is not like this. It first establishes the project, and then all participants share the token. The initial participants didn't know whether the tokens they got were worth or not.

In fact, the value of Token depends on the number of participants, whether they are actively involved, and whether they bring their intelligence to the project. If more and more people are involved and everyone is very active, while all contributing a lot to the project, then the token is valuable. This is a huge innovation in itself.

It's like a few decades ago when rural production teams kept work points. A young worker was credited with 10 work points for a day's work. I was a primary school student at that time, and I was credited with one work point a day for helping out on the farm during the busy season. We didn't know how much the work points are worth, but at the end of the year, the production team would calculate how much rice and how many pigs and cows our production team has sold this year, and how much money we earned.

For example, the income is 100,000 yuan, and the entire production team has accumulated 10,000 work points this year. Then the production team will hold a meeting and announce that one work point is worth 10 yuan this year.

Unlike the pre-valuation of internet startups, you do a thing on the blockchain and I give you a Token, how much is it really worth when I give it to you? It's actually uncertain. You only know after you've done something. So this fundamentally changes entrepreneurship and brings the threshold of entrepreneurship down to the minimum.

Let's see Uniswap again. How much start-up capital do we need to write 500 lines of code? There is no need for Angel investment at all. What is the current market value? Nearly $30billion. Where did that $30 billion come from? It was brought by everyone. The 500 lines of code just write a transaction contract, and then all participants use this contract to make transactions, and then this project is worthy.

So I often joke that there is only one condition for blockchain-based startups like Uniswap, and that is if we can rely on our parents, then we can start a business because all we have to do is to stay home and write 500 lines of code. If no one uses it, it doesn't matter. It's only 500 lines of code anyway and we can write another one. What's the big deal?

Tradewind Think Tank: How does the token economic model link and motivate stakeholders?

Xiao Feng: That's the change token brings, turning shareholders, in the traditional sense, into stakeholders in a broader sense. It is not just friendly to shareholders. The term "maximum shareholder interests" coined in the 1970s is completely invalid in the blockchain era.

By emphasizing the maximization of shareholders' interests, the operation of CEOs of listed companies will become shorter and shorter because they have to report every month, — report earnings and revenues every quarter, which has to show the growth. This is maximizing shareholder interests.

In the Internet era, there are actually many entrepreneurs who claim not to maximize shareholder interests, but to put their employees first and shareholders last. An over-emphasis on maximizing shareholder interests is not helpful to the long-term value of a company and can be damaging.

In the blockchain era, most projects do not have shareholders, but a large number of stakeholders who hold tokens, and when this project is done well, they all have an incentive in terms of interests, so everyone is willing to participate and invest. This is the change brought about by the token economy.

Of course, everything has two sides. Under the guise of blockchain, some unscrupulous institutions conduct illegal financing, pyramid schemes and fraud, which disrupts the social and economic order and hinders the applications and development of the blockchain itself. In this regard, we should first improve the regulatory system from the level of policies and regulations, and strengthen the crackdown. Furthermore, blockchain practitioners should strictly abide by laws and regulations, legally use blockchain, and develop innovative applications based on real scenarios. Besides, the public should also be rational and cautious to judge according to laws and regulations, and enhance their awareness of financial risks.

Tradewind Think Tank: For practitioners in the blockchain, how can they find application scenarios suitable for blockchain to fully play their value and what are the common features of these scenarios?

Xiao Feng: We must first have a core concept that blockchain cannot be regarded only as a tool or a technology. If we really want to use blockchain to start a business, we have to think clearly about what we want to change with blockchain. Because blockchain is not only a technical tool, but also an economic model. This requires us to design a sound game mechanism and business model when we innovatively apply blockchain. So, the innovative application of blockchain in the real sense is a very complicated thing.

I have met many people who do very well in the Internet, and it has become a lot of work for them to understand blockchain. This is because the logic of blockchain is completely reversed, or another set of logic. The more successful they are in the internet space, the more difficult it is for them to jump to the blockchain space.

Therefore, we must understand that blockchain is comprehensive. It has not invented any new technology by itself, but it has integrated numerous technologies emerged in the past few decades. It made a fusion innovation and formed a very sophisticated blockchain system.

At the same time, we must also be particularly careful not to use blockchain for the sake of blockchain and not to decentralize for the sake of decentralization. If the problem can be better solved by other methods, such as using a centralized method to better confirm transactions, then make sure not to go for blockchain.

We must find a problem that can only be optimally solved by blockchain or decentralizing. Only then is it likely to work. For example, Uniswap can achieve a daily transaction volume of $7 billion with only 11 people and 500 lines of code. Is there anything better than that? Is there anything lighter, cheaper and more efficient than it? The answer is "no" and of course it will succeed.

Blockchain Naturally Matches the Market Circulation of Data Elements

Tradewind Think Tank: How does the asymmetric encryption, distributed ledger and, consensus mechanism of the blockchain play a role in data credibility and security?

Xiao Feng: Here are two problems. Firstly, whether the collected data is reliable or trustworthy. Our current model is guaranteed by a centralized organization, or in a centralized way. Whether it is a big data bureau or a certain company, the authenticity and reliability of the collected data are guaranteed by this way.

But in the future, when 100 billion devices are connected to the Internet, is this still the best way to ensure data authenticity and reliability? At that time, using centralized institutions and methods will be a costly and commercially unviable model.

For the second question, let's take V2X as an example. How do we ensure that the two devices immediately complete the data exchange on site within a fraction of a second? How do we know no device has been hijacked by hackers? And we often see in the news reporting that hackers have hijacked IoT devices in western countries.

Therefore, we must first solve the problem of data credibility. Blockchain is by far the best, the cheapest and the most suitable solution. Because it is impossible for one centralized organization to ensure the credibility for such a huge number of devices. Even it is possible, how large a team or centralized organization we need to ensure data credibility? So, at that time, we must make all the data to be generated on the blockchain.

The asymmetric encryption of the blockchain is known to us all. And so far, no one has deciphered the private key on the blockchain, as it is very difficult for hackers to attack the data on the blockchain. The data on the blockchain is public to all users so it is invalid for individuals to modify the ledger. And the cost of colluding with so many nodes is so high that it is technically guaranteed to go from "no hacking" to "unable to hack."

Blockchain is a network that can't be hacked, so as long as data occurs on the blockchain, we can completely trust it because it cannot be changed or deleted.

How is the data generated on the blockchain? This depends on whether IoT is based on the Internet or blockchain. Data based on the Internet may be hacked by hackers or modified by the owner of the device. But if based on the blockchain, the sensors of the Internet of Things are written on the blockchain, the data is sent through the blockchain, and it is impossible to be hacked or modified.

Tradewind Think Tank: What role related technologies such as blockchain and privacy computing play in data assetization and market circulation of data elements?

Xiao Feng: Data assetization, data circulation and blockchain are naturally matched. There are three problems. First, we must ensure that the data is true, otherwise how can big data be credible? How can the artificial intelligence algorithm be trained? As we mentioned earlier, this problem can be solved by generating data on the blockchain.

Second is the confirmation of data ownership. Of course, it is fine to use existing centralize method to determine the ownership. But let's think about the future. When 100 billion devices are sending and receiving data, how can we use a centralized method to determine data's ownership? How to determine who generated it? How should we distribute the benefits, or what reward one should get after contributing data? At that time centralized would be extremely expensive.

Therefore, blockchain is used to confirm the rights with value tools like NFT to mark the uniqueness of data. Obviously, blockchain is a good tool for data confirmation.

Data confirmation is not the ultimate goal. Because data is only valuable in exchanging. Whether it's your data or my data, we will always have data, but if we don't exchange, the data waste. Before data can be exchanged, the third issue must be addressed, that is privacy protection.

This is where privacy computing comes in. In the crypto state, we can create a desired result that the data is "available but invisible" during the exchange. At the same time, there is also the issue in clearing and settlement. The blockchain is essentially a new type of clearing network. In addition to the clearing of digital currency exchanges, the exchange of data also needs to be settled. So, the blockchain itself can also help us to do clearing in data exchange.

Therefore, in terms of credibility, confirmation, and circulation, blockchain and big data are perfectly matched.

Use Digital Currency to Serve the Decentralized World

Tradewind Think Tank: You mentioned that China launched sovereign digital currency issued by the central bank. What will be the practical implications of it? What will be the impact?

Xiao Feng: When everything is connected, we may need a decentralized technical architecture in many aspects to support such a complex network with such a large amount of data communication and huge resource management, so that it will be cost-effective. Then, on this basis, whether in economy or business, they all need to be decentralized. Because the base is decentralized, it is impossible to solve economic and commercial problems in a centralized way, which is also not cost-effective.

At this time, both the sovereign digital currency issued by the central bank and the crypto digital currency on the blockchain, are actually designed to correspond to this decentralized world and distributed technology, economy and commerce.

There is actually no need to design digital currencies if we are not going to deal with this situation, that is my opinion. Because if we don't have to think about how to manage the IoT and how to govern the decentralized world, our current electronic currencies would be sufficient, especially in China, 100 per cent sufficient. We already have a good electronic payment network.

So, I don't think digital currency is a duplication of existing electronic money, but because we want to govern a centralized world after everything is connected, we need to reconstruct a new generation of digital financial system based on blockchain to match the decentralized world.

Digital currencies were created for this purpose, otherwise they would be meaningless. Whether it is a sovereign digital currency or a crypto digital currency, the ultimate value lies in governing this decentralized world, not in making a new electronic payment system.

The essential cause of this decentralized world is the interconnection of everything. 100 billion devices need to be end-to-end and decentralized to make decisions, predictions and many other things. For example, a lot of predictions and decisions must be made when driving a car, which involves complex and huge data collection, exchange and calculation.

Digital currency serves this purpose. In a decentralized world, it serves to the decentralized production and life by using coin, token and NFT. This coin can be a sovereign digital currency or a crypto digital currency.

Tradewind Think Tank: Could you please project for us how digital currencies will gradually penetrate our ordinary production and life?

Xiao Feng: It depends on the digitalization of the economy, the digitalization of life, the digitalization of cities, and the digitalization of enterprises. But the COVID-19 has accelerated digitalization process. Coupled with the development of 5G and Internet of Things technology, it will further accelerate the digitalization of the entire society and economy.

Therefore, we can predict that a decentralized world will surely emerge in the next decade. When this kind of change reaches a certain level, it will prompt us to establish and formulate supporting technologies, mechanisms, models and laws. The faster the decentralized world grows, the faster this process will be.

We can already see a lot of innovations, which is actually serving the fast growing and changing decentralized world.

In fact, the blockchain technology itself is also developing. Till today, after more than 10 years, the infrastructure of blockchain is merely completed. For example, the new technologies such as Filecoin, Dfinity, OSMOS, and Polkadot have just emerged in recent years.

Then how to store huge amounts of data in a decentralized world? There must be decentralized storage, which is what Filecoin does. After collecting so much data, how to calculate it? Send all to the cloud computing room? That's not realistic. So, technology like Dfinity appeared. It is called an Internet computer, without centralized computer room and calculating on the terminal node based on the blockchain. This is decentralized cloud computing.

In a decentralized world, there are various blockchains, which are isolated from each other and cannot communicate. Therefore, we need to solve the problem of cross-chain and multi-chain communication. COSMOS mainly solves the problem of cross-chain communication, allowing communication and transactions between various blockchains. And Polkadot mainly solves the problem of multi-chain integration, integrating multiple blockchains into a unified blockchain network.

Coupled with the financial infrastructure provided by more well-known crypto digital currency, so far, the overall technical framework based on the blockchain has been completed.

On this basis, we still need 3 years, 5 years or even 10 years to optimize the entire system. Construction is one thing, optimization is another. Construction can rely on creativity, but optimization depends on specific uses. Just like an artificial intelligence algorithm, the algorithm can be written by a scientist like writing a paper, but it can only be checked and optimized in a certain specific situation.

But this is not to say that specific innovative applications will only emerge in 10 years. They will emerge continuously and gradually through time. Because the threshold of blockchain-based entrepreneurship is already very low.

The threshold for starting a business is such low that it can be done with 11 people writing 500 lines of code. What does it mean? It means that there were 1 million entrepreneurs in the past, but now the number can become 10 million. In the past, I cannot start a business without enough money and resources. Even if I have some good ideas, I cannot overcome the financial and technology barrier. But now the blockchain has sunk everything into an on-chain infrastructure. Based on the above services, only 500 lines of code are needed, the technical threshold has also been lowered, and the capital threshold has also been lowered. This means that there will be more and more entrepreneurs. So, let's image how prosperous the next 10 years will be.

Sylvain Laurent:
How to Promote the Integration of the Digital Space and the Real World

What is the nature of the digital twins? How should we properly understand the digital twins? Can digital twins significantly reduce the cost of trial and error in the real world? What is the purpose of developing digital twins? What role will digital twins play in "carbon neutrality?" What future advances will the digital twins bring us in the future?

Use Digital Twin Technology to Reduce Trial-and-Error Costs

Tradewind Think Tank: Dassault Systèmes has been dedicated to the field of digital twins for a long time. Could you please tell us what is the nature of digital twins? How should we correctly understand digital twins?

Sylvain Laurent: As you said, the concept of digital twins has existed for a long time. For Dassault Systèmes, the application of digital twins was initially in the aerospace field. We made digital prototypes for aircrafts.

Relying on digital prototypes, we have spent a long time exploring how to use a virtual world to improve the development of the aerospace industry since 1990. That is the starting point of our Dassault Systèmes research on digital twins.

In the last five years, we have developed from the concept of digital twins to virtual twins.

The difference between virtual twins and digital twins is that the former has evolved from a single digital prototype to a compositive prototype which includes relevant information and functions, making it more abundant.

In fact, a virtual twin is the simulation of the real world, and its concept is much broader than 3D modeling itself. We have applied the concept of virtual twins to all walks of life.

Six years ago, Dassault Systèmes' virtual Singapore project promoted the widespread application of our virtual twin technology. In fact, all walks of life now need a particularly realistic virtual world that can reflect the real world more truly. For us, virtual twins are the only effective way to reflect.

Tradewind Think Tank: What effect will a digital twin world or object exert on the physical world or objects, and what are the benefits?

Sylvain Laurent: The answer is obvious. Nowadays, every system in the world is becoming more complicated, from all industries to every product and even the human body is a very complicated system.

Take our virtual Singapore project as an example. In fact, Singapore's cities were built from scratch. After 50 years' development, Singapore has made great achievement. However, in the next 50 years, what kind of development strategy should

Singapore adopt? Now Singapore's leaders need to use virtual twins to manage the complexity of the city, so as to help them make better decisions, such as how to build more well-found infrastructure and how to predict and deal with public safety issues.

Another example is about the automobile industry. In the past, if automakers wanted to develop a new model, they needed a lot of test cars to test parameters like acceleration, wind resistance, safety, and collisions. After extreme tests, most of these test cars became scrap and must be destroyed, which caused not only the high cost, but also a waste of time.

Now, 90% of our simulations and simulation tests can be completed by using virtual twins, and this process will be safer. Also, different parameters can be selected for debugging according to the different performance of the vehicle. As a result, we can use virtual twins to develop a new generation of models, including future driverless cars and connected cars.

It is safe to say that the virtual twin is not a simple 3D modeling, and it connects and reflects the function and performance of the object. The object can be a car, a human being, a city, an airplane or even a factory. More importantly, the virtual twin not only thinks about the object itself, but also takes the surroundings into consideration, and analyzes how the object and the environmental factors affect each other.

Tradewind Think Tank: The biggest cost of innovation is trial and error. Can we say that digital twins can greatly reduce our costs of trial and error?

Sylvain Laurent: You get it exactly. There is no rehearsal in real life, but virtual twins give us the opportunity to see what will happen in advance. Digital twins or virtual twins can help us shorten the entire product development cycle. Besides, it can provide us with a variety of solutions. If we need to compare among multiple solutions or multiple scenarios, it can help us reduce costs and risks. For the management, it can also speed up the efficiency of decision-making, which are difficult to achieve by using actual objects and entities.

In addition to shortening time and reducing costs, as what is emphasized above, we can also place virtual twins in various real-life scenarios for observation. For example, in a city scenario, you can put a car on a road, and then observe the behavior of the car, including what requirements it should meet to achieve a comfortable and safe driving state in different situations. That is very convenient for digital twins.

For example, we can perform simulation tests through digital twins to see different performances of the same care in Beijing, Guizhou, and Xiamen. The driving performance of the same car must be different from other cities because Xiamen is a coastal city and is warmer than other cities. Also, users in different areas have different expectations for cars, so we can use virtual twins to simulate in different scenarios.

A Bridge Connecting the Real Economy and the Digital Economy

Tradewind Think Tank: Digital twin is more acceptable because of its simple and intuitive experience, so can it help to accelerate the implementation of various digital applications?

Sylvain Laurent: I agree with you. Currently, digital economy needs reliable simulation technologies as support. Whether it provides a variety of products or solutions, we all need digital twins and virtual twins to promote the integration of digital applications and actual scenarios.

At Dassault Systèmes, we call this new economic situation the experience economy. We believe that in the future, we need to provide a variety of users with digital experience to meet their expectations.

So I think we must use digital twins and virtual twins technology to boost the development of the digital economy.

Tradewind Think Tank: Basically speaking, what is the purpose of our development of digital twins?

Sylvain Laurent: Whether virtual economy or virtual twins, a very important purpose is to manage complexity, including local complexity and global complexity. The virtual world can provide a simple way to manage the complexity because this complexity includes a variety of disciplines and it also requires a combination of various functions.

Tradewind Think Tank: The integration of the digital economy and the real economy is the trend of the times. Do you think the digital twin is a bridge connecting the digital economy and the real economy?

Sylvain Laurent: Your understanding is totally correct. We believe that virtual twins can be used to connect real and virtual economy as well as the two forms of the world in a simpler and more convenient way. Of course, it is still challenging, and we have to extend trust to the digital model of the virtual twins.

Everyone may have different expectations and needs for the digital model of the virtual twin, so it must integrate a variety of different information. In fact, virtual twins are more like a new language. Using virtual twins, we can manage complex data, information, elements and models, which can be shown in a very simple way at different levels and in all aspects, thereby reducing chances of making mistakes.

Data Integration Is Behind the Digital Twin

Tradewind Think Tank: Based on your experience with Dassault Systèmes, what are the key points to build a successful digital twin application?

Sylvain Laurent: Above all, we should talk about methodology including what kind of methodology do we need to describe the application scenarios of virtual twins. In Dassault Systèmes, we have such a methodology, which is value-oriented. There are a lot of industry consultants in Dassault Systèmes who have extensive experience of all walks of life. They practice our value-oriented methodology through the following four steps to provide service and support for customers.

In fact, the four steps are easy to understand. The first step is to value evaluation such as evaluate the current state of the situation and figure out the current situation. The second is to define the value including define what kind of value is to be provided, that is, what the scenario needs. The third step is to promise the value. We promise our customers what value we will deliver. The fourth step is to deliver the value that we promise.

If we say that we do not use this set of methodology, we may not be unable to achieve a digital twin or virtual twin application. If so, such a virtual twin will be short of a more accurate definition to define what goals it should achieve. We can use virtual twins to help people understand what value we can deliver and what goals we can achieve.

Tradewind Think Tank: Dassault Systèmes proposed a 3D experience strategy as early as 2012, and then launched a 3D experience platform. Could you please tell us how Dassault Systèmes serves and empowers global users?

Sylvain Laurent: Just now we mentioned that virtual twins can be used to manage complex and diverse information and data, and can also be used to manage employees, institutions, and ecosystems. The 3D experience platform we launched essentially is a digital platform that can help us connect the digital world and the real world, and at the same time can breakdown data silos and connect all data together.

We can use a new concept, the 3D experience platform, to gather people from different regions, countries and industries in the same virtual space, discuss topics of common interest, and coordinate to complete a goal.

For example, in the past, if we need to build a 3D model of mechanical parts, various departments had to communicate through emails or file packages. This process was inefficient as misunderstandings and deviations were prone to occur. In fact, if all relevant employees in different departments, regions, and supply chains are gathered on the 3D experience platform, and everyone deals with the same 3D model of mechanical parts, they can communicate and work together vividly and efficiently.

What's more, many interdisciplinary and cross-industry virtual twin models have been produced in this process, such as urban architecture, biomedicine, and automobile machinery models. The core technologies are common. These models can share data with more organizations on the platform that have the same needs. While improving everyone's efficiency, it can also make the model itself more valuable.

The 3D experience platform can be used to communicate among different groups of people, and it also can manage a large amount of data. The core of digital twins is not about a paragraph or a sentence. In fact, it carries a variety of different information and data, and manages massive amounts of information together. In addition, the platform is able to solve security issues, including data security, and in which country, or in which scenario, who has access to use the data.

Why do we want to name the platform 3D experience platform? The reason is that we realize various objects, and our human body are all three-dimensional, and our idea uses virtual three-dimensional models to describe, map and experience the real world. That is what we have just mentioned. We have now entered an era of experience economy, and we can try various possibilities through virtual twins.

Tradewind Think Tank: You mentioned the integration and management of data. Are we easily attracted by the superficially cool 3D images, but neglect the integration and application behind the data that are really important?

Sylvain Laurent: Let's go back to the virtual twin you just mentioned. The virtual twin connected to the 3D experience platform can be seen as an iceberg. What we see is only the tip of the iceberg. The truly large amount of valuable data is hidden under the sea and is invisible.

We just said that we can use such a virtual 3D experience platform to manage complex issues in a simple way. For example, in the manufacturing field, we can use virtual twins to integrate all relevant information and data in the entire value chain, so that engineers can coordinate their work more efficiently and conveniently. That is also can be applied to all walks of life.

Tradewind Think Tank: Would you please share the changes that digital twins brought to all walks of life?

Sylvain Laurent: We notice that virtual twins play an important role in all walks of life, no matter in the past or in the future. In the aerospace industry, if there were no virtual twin technology, it would be difficult for us to efficiently build an aircraft, and would take a long time to design and build an aircraft because the aircraft design and manufacturing process require a variety of different simulation tests, including the application of new materials and various new technical applications. Now, we can shorten the time from 20 years to 4 years.

Let's talk about the automobile industry. In terms of new modes of transportation, our concept of a car is not only about how its body, chassis, or engine should be designed, but also about the interconnection between the inside and outside. We should drive the car in an application environment, do some tests and observe its performance, no matter in a city or on an intelligent road. If there were no virtual twin, developing the next generation of cars would definitely be more troublesome and would take longer time. Using virtual twins can greatly save time and energy.

What we mentioned are some examples of manufacturing. In fact, there are many examples in all aspects of infrastructure, smart cities, and life sciences. I would like to talk about the goal of carbon neutrality that is being globally emphasized, especially in China.

In fact, when we design or build a product, we always consider its impact on carbon emissions. We can use virtual twins to make accurate predictions about the impact of carbon emissions in the manufacturing process and even in the logistics process.

What we talk about is to bring value to all walks of life and help them to innovate. Take the construction industry as an example. The current construction industry is undergoing a new round of changes. In the future, there will be a revolution in the application of new materials and new construction methods, including new factory construction methods. In the next five years, construction companies of all sizes will transform into digital.

Let's talk about the life science industry. The COVID-19 pandemic has raged around the world which is a disaster for all of us. We can notice that the life science industry uses digital technology to speed up the development of vaccines. Previously, the vaccine development cycle was about 10 years, but now we have developed a vaccine within two years.

The technology behind that is virtual twins. It has simulated a variety of different macromolecular institutions. Based on algorithm models, three or four combinations of macromolecular structures with the most pharmacological effects and the lowest toxicology are found from more than 1 million combination possibilities.

Digital Twins Make Cities Better

Tradewind Think Tank: You just talked about "carbon neutrality". This is a topic that China and the world are paying great attention to, and it is also a goal that human beings will achieve in the future. What role can digital twins play in the field?

Sylvain Laurentn: In terms of "carbon neutrality," virtual twins mainly play a role in statistics, monitoring and prediction. There are too many application cases. For example, for a city, based on the current production and living standards, decision makers want to know how much total carbon emissions would be and how to achieve refined statistics?

The sources of carbon dioxide emissions include stationary factory chimneys and buildings, as well as vehicles such as airplanes and cars. So we need to use virtual twins to build a model to manage the complexity. For example, count how many coal-generated power plants, how many factories, how many people and cars there are in the city, and then refine them to each regional cell, each district or park, and even each building. In this way, it is clear to know which areas in the city are the high carbon emission areas and the density of emissions, which can also be marked with different colors.

The next step is to solve it. If we find that the biggest part of carbon emissions is coal-generated power plants, how should we manage it? Is it a direct or a gradual shutdown, or an energy transformation? This is also a dynamic decision. Therefore, we need to build a prediction model based on virtual twins and use data to deduce. If policymakers require a 50% reduction in carbon emissions within a year, does it mean that most power plants have to close, which will cause many people to lose their jobs? What effects will it have on the whole industrial chain, and how much impact will it have on the economy? Based on the comprehensive analysis of those related factors, we will then deduce the choice, whether to shut down gradually in three years and five steps, or five years and three steps, or replace it with natural gas, solar energy, and geothermal, to determine which method is the best decision.

Tradewind Think Tank: From the perspective of digital twins, what do you think are the global misunderstandings of the perception and practice of smart cities?

Sylvain Laurent: In fact, in the last decade, we have found that many people all over the world were discussing the concept of smart cities, but few have really put it into practice. Why is this so? The reason lies in the lack of a digital twin model of a city.

Currently, most government management departments are still using piles of paper. Although various departments have innovative ideas and applications, they are rarely brought together in an effective way.

When it comes to the smart city project of virtual Singapore, we are using virtual twins to achieve data interconnection. This is not just a cool-looking 3D model, it is a true digital twin, helping policymakers make decisions faster and more accurately through data. For example, in terms of energy utilization, urban security and infrastructure, we can simulate and predict whether vehicles will be congested on the road, whether it is safe for residents to cross the road without affecting traffic efficiency, and how much power supply should be allocated in a certain area to support future development.

Using digital twins in the process of city construction and operation, integrating various actual conditions and possibilities, and making optimal decisions based on data analysis is an important achievement for Singapore to build a smart city. We also plan to implement such projects in some cities in China to help decision-makers make correct decisions faster.

In fact, collecting real-world data is no longer a particularly big challenge. We can use the Internet of Things for data collection, but it is difficult to predict future changes which will affect citizens. That is the core of virtual twins. Virtual twins have this ability to help decision makers observe and analyze data, and then make predictions about the impact of the decision in the future.

China's Development and Future Trends of Digital Twins

Tradewind Think Tank: In your opinion, what advantages does China have in developing digital twin applications? How does Dassault Systèmes deploy in the Chinese market?

Sylvain Laurent: I think China is developing very fast in all walks of life. Besides, we need to accelerate the integration of the digital economy and the real economy, and the application of digital technology, so that all walks of life can move forward at the same pace.

In the China's market, Dassault Systèmes can provide support including using virtual twins and digital twins to empower all walks of life. We have learned about China's "14th Five-Year Plan," and the development goals of Dassault Systèmes are highly consistent with it. For example, the "14th Five-Year Plan" sets goals such as carbon neutrality and cultivating the next generation of high-tech talents. Those goals are highly consistent with our goals.

In fact, Dassault Systèmes has already applied in many well-known projects in China. For example, the Bird's Nest National Stadium, the new CCTV building, Daxing Airport, the domestically produced large aircraft C919, and the world's first paperless construction ship Haixun 160.

In recent years, we have seen an increasing number of original innovations in China, which are often without reference, leading to great uncertainty. In the process of reducing uncertainty and the trial and error cost of innovation, virtual twins are very useful. For example, we assist China's aerospace industry by simulating the design, construction, launch and operation of rockets and satellites, assist the manufacturing of China's large ships and high-speed railways, and cooperate with Chinese new energy vehicle companies such as NIO, XPEV, and Li Auto.

In order to expand China's market, we recently established an intelligent manufacturing innovation center in Chongqing. We also hope to use our virtual twin technology to empower Chongqing's manufacturing industry and help Chongqing build a smart manufacturing town.

In addition, during the epidemic prevention and control, one thing that impressed me a lot is that in the process of constructing Leishenshan Hospital, we cooperated with CSADI and donated software to help them discuss and formulate conventional HVAC installations based on the characteristics of on-site implementation.

We also carried out fluid mechanics simulation calculation, and finally confirmed the solution for the best control of the pollutant concentration, and the timely delivery of the patient's exhaled pollutants to the exhaust outlet based on the calculation results. At the same time, according to the simulation results, we can also put forward some suggestions on the location and operation of the medical staff on site. These not only minimized cross-infection of medical staff, but also reduced impacts on the community outside the hospital and the surrounding environment.

Finally, we believe that it is very important to cultivate the next generation of high-tech talents. Education is actually a very long process, and we hope to empower schools, and to small-and-medium-sized enterprises, so that the next generation of talents can be exposed faster to such a digital world, and innovative SMEs are encouraged to develop their ecosystems.

Tradewind Think Tank: Digital twins are more widely used, and surgical operations can be simulated by digital twins. Would you mind telling us what else can the digital twin bring to us in the future?

Sylvain Laurent: Actually, the 3D digital world we live in is developing rapidly, especially in China. I believe that maybe we can help China's space industry to explore Mars together.

For the future of digital twins and virtual twins, I want to make three points. The first is the surgical operation you just mentioned. In fact, I think that in the next 10 years, we can better predict and control the human body. We can compare the human body to a complex system like an airplane which consists of a lot of data. We can use virtual twin technology to predict what kind of disease the patient may have before undergoing surgery, which is able to avoid the deterioration of the disease.

In this regard, we have also done a very sensational thing, which is to create an artificial heart through virtual twins. In fact, we have been studying a long time how to use the so-called cloning technology to create human organs, thereby prolonging human life, or solving many diseases that are currently incurable.

One of our most remarkable results is that an artificial heart was cultivated from real human stem cells, and then a heart transplant was performed, which prolonged the survival time of patients. This is an advancement in human science.

Behind this, we built a virtual twin model of the heart, including all the details of the overall structure of the heart, muscle walls, blood vessels and nerves, and truly restored the human heart organs one by one. The applications extended from this are very valuable. For example, when a surgeon is faced with a relatively high-risk heart operation, the heart may be very fragile and prone to explosions and other risks. In this case, the surgeon can use the virtual twin heart to synthesize blood pressure, heart rate, blood viscosity, and other physical signs data, repeatedly perform surgical simulations, repeatedly try and error to eliminate risk points, and then develop the most reliable and safe surgical practice. The operation method greatly improves the success rate and safety of the operation.

The second point is the next generation of infrastructure construction and urban management. For example, Shanghai and Chongqing are both large cities. How to help the government make better decisions for such a large city will also rely on virtual twin technology.

The third point is the goal of carbon neutrality. I think virtual twin technology can be used to protect the earth's environment no matter it is now or in the future.

Li Yuanqing:

How Brain-Computer Intelligence Nurtures the "Most Powerful Brain"

How to understand brain science? What are the new meanings given to brain science in the era of intelligence? What stage and level has the development of brain science researched in countries around the world? What is China's "one body, two wings" brain program, and how are the "one body, two wings" related and synergistic? What will be the main focus of the next industrial application of brain-computer interface and what problems still need to be solved?

From "Understanding the Brain" to "Protecting the Brain" and "Creating the Brain"

Tradewind Think Tank: Now that intelligence is booming, how should we understand "brain science" and what new meanings has "brain science" endowed in the era of intelligence?

Li Yuanqing: A very important goal of artificial intelligence is to achieve general artificial intelligence, which is very similar to the structural principle of the "human brain." The essence of the brain is to process internal and external information, such as internal experience and knowledge, and external perception, and then make corresponding responses and decisions. In the same way, artificial intelligence should be like the brain, processing non-standardized information in nature, making automated responses and intelligent decision-making, not just in statistics, and playing games, chess, or super memory. From this perspective, the development of artificial intelligence and "brain science" is complementary and mutually reinforcing.

On the one hand, the development of brain science can provide new methods for artificial intelligence. The more we understand the logic of the brain's operation, the more we can integrate various fragmented AI technologies, thus guiding AI models, algorithms and products to further enhance and create more intelligent information processing systems, or even reach the capabilities of the human brain. This is the significance of brain science to promote the development of artificial intelligence.

On the other hand, new technologies and methods of artificial intelligence nurture the research of brain science. All the activities of human society are the result of the operation of the brain, so brain science has been trying to understand how the brain works, such as how the brain makes decisions. What is consciousness? Why is there innovation? Why are there different personalities? Artificial intelligence can not only collect the signals and data of the brain through the BCI technology, but also analyze and decode the brain data through the algorithm model, so as to help people further understand the operational logic and decision-making mechanism of the brain and boost the research of "brain science."

In fact, humans have made some breakthroughs in brain research, especially in the diagnosis and treatment of some brain diseases. For example, early diagnosis or intervention treatment of these brain diseases can be realized through artificial intelligence.

Tradewind Think Tank: How is the "brain science" research going around the world?

Li Yuanqing: There is no doubt that brain science will have a profound impact on the development of society in the future, and has become a strategic high ground for global competition in science and technology.

In 2013, the White House proposed the "Brain Project," which aims to explore the working mechanism of the human brain, draw a full picture of brain activity, promote neuroscience research, and develop new therapies for brain diseases that are currently incurable. In the same year, the European Union gathered more than 400 researchers from different fields to set up a special project — the "Human Brain Project (HBP)." This project hopes to simulate the brain through computer technology and establish a new and revolutionary ICT platform for generating, analyzing, integrating and simulating data to facilitate the application of research results. Japan's "Brain Project" focuses on the medical field, mainly using the marmoset brain as a model to accelerate the research of human brain diseases, such as Alzheimer's and schizophrenia.

China has included "brain science and brain-like research" as a major national scientific and technological innovation and engineering project, and proposed a "one body, two wings" layout plan. At the beginning of 2021, the Chinese Academy of Sciences established a brain science and intelligent technology innovation center with 20 institutions and 80 elite laboratories, dedicated to the in-depth study of early diagnosis and intervention of brain diseases, brain science and brain-like intelligent devices. Compared with European and American countries, China's "brain science" research started relatively late, but it has grown rapidly based on a solid scientific research.

The core of BCI technology is to collect brain signals, decode and process them, and then generate a series of instructions to realize the direct interaction between the brain and external devices. At present, we are trying to combine BCI technology with various traditional scenarios, hoping to solve problems in multiple fields such as medical care, culture, and safety.

Tradewind Think Tank: What does "one body, two wings" proposed by China refer to, and how are the "one body, two wings" related and coordinated?

Li Yuanqing: "One body" means to continuously recognize and understand the brain. The inside of the brain is a vast "universe" with too many unsolved mysteries. If we can better understand the logic of the brain's operation, we can better understand the whole world. For example, the "Atlas of Mesoscopic Neural Connections of the Whole Brain," the international science project, hopes to map the connections of brain neurons and cells to help us better understand the brain. As we analyze a chip of a computer, we must figure out the wiring connections inside to know how it processes information. This project is a typical research of "understanding the brain".

The second is the "two wings". On the basis of understanding the nature of the brain, we can better protect the brain, simulate the brain and even create the brain.

One wing is to protect the brain, which studies the diagnosis and treatment of brain diseases so as to form a variety of new medical industries. In terms of how to protect the brain, preventing the brain from declining and brain diseases is very important to humans. Another wing is to simulate the brain or create the brain, which focuses on the research of new technologies related to artificial intelligence, such as brain-like artificial intelligence, brain-like computing, and BCIs, aiming to create machines that are as intelligent as humans. This is an important goal for the development of artificial intelligence in the future.

The core of "one body, two wings" is "understanding the brain," "protecting the brain," and "creating the brain," which also corresponds to the three major research directions of "brain science." "One body" is the cornerstone, while "two wings" are development and extension. Only deeply recognizing and understanding the brain can we improve the function of the brain, protect the brain, and create the brain, and they are complementary to each other.

How Artificial Intelligence Learns the Human Brain

Tradewind Think Tank: What constitutes the comprehensive brain science?

Li Yuanqing: Brain science is a broad and extremely complex research. In a narrow sense, brain science is neuroscience, which studies the difference of the nervous system at the molecular level, cellular level, and between cells, as well as the integration of these processes in the central functional control system. American neuroscience defines brain science in a broad sense, and considers it to be a science that studies the structure and function of the brain, and cognitive neuroscience. In general, brain science is a highly cross-integrated major, requiring background knowledge in fields such as mathematics, physics and chemistry, computers, informatics, psychology, biology, and medicine. In the entire brain science research, these disciplines are intersecting, synergizing, merging and supporting each other. For example, although medicine mainly focuses on the research of "protecting the brain", medicine itself has a large amount of medical data related to the brain, which can provide basic data support for the research of "creating the brain."

In fact, the research and development of brain science requires to integrate various disciplines, and supports the application scenarios so as to better promote the research of brain science, and to bring benefits for mankind with brain science research findings. For example, the face recognition technology of artificial intelligence is derived from the "cell assembly hypothesis" put forward by Canadian psychologist Donald Olding Hebb.

It can be summarized simply as: when a group of nerve cells in different parts of the brain stores a memory, given a partial memory stimulus, all the nerve cells in this assembly will be activated to achieve contrast recognition and re-memory.

How to Connect Human Brain and Computer

Tradewind Think Tank: When brainwaves change from brain signals to electrical signals, what is the logic principle behind BCIs and what are the key technical points?

Li Yuanqing: BCI, namely, is "brain+machine+interface," which enables the brain and machine devices to directly establish a new information channel through interface. The working principle of BCI can be divided into four steps: signal acquisition, signal processing, control equipment and information feedback. Specifically, the BCI technology collects Electroencephalogram (EEG) signals from the cerebral cortex through signal acquisition devices, converts them into signals that can be recognized by computers after processing, and then extracts the characteristic signals for pattern recognition, and finally converts them into specific instructions to control the external devices.

Therefore, the essence of the BCI is to establish a brand-new communication technology and control technology between the human brain and the real world. For example, human behavior is made by the brain to give instructions, and then the body organs receive signals to respond, such as walking, running and playing ball and other sports. If the limbs of the human body are severely damaged, they will lose their motor functions. In this case, as long as the brain is still functioning normally, we can equip the patient with a robotic prosthesis, and then use the BCI technology to realize the brain-controlled prosthesis, thereby helping the patient to recover their exercise ability.

What are the key points of BCI technology? The first is the collection of brain information. We know that the core functional area of the brain is in the cortex. The various parts of the cortex are responsible for processing auditory, visual, and various sensory information. At the same time, they are also in charge of many aspects such as language, movement, thinking, planning, and personality. The information collection of the BCI is mainly for the extraction of the inner cerebral cortex under the skull, so the interface method and technology are particularly critical. It not only needs to obtain high-quality brain signals, but also ensures that the brain itself is not damaged.

The second is feedback. An important part of brain thinking and decision-making is getting feedback. Humans perceive the environment through perceptual ability, and transmit perceptual information to the brain for feedback, such as whether the object is hard or soft, hot or cold, red or green, and the brain can proceed with the next processing and decision-making through the feedback information. This process is simple for human perception, but it is actually extremely complicated for BCI. Since the BCI needs to feed back the environmental information obtained by external machines and equipment to the brain, its path and mechanism are completely different from human perception, and it needs to use hybrid resolution technology of multimodal perception.

To give a simple example, although the external machine can perceive an object, it is difficult to distinguish the physical properties of the object, such as "soft" and "hard." In addition, physical properties are also related. When an object has both "hard" and "smooth" properties, it will often also have "cold" properties, which is the synaesthesia between hardness and temperature. Such complex tactile attributes are extremely difficult for technical perception. The hybrid resolution technology of multimodal perception perceives the physical properties of objects by constructing a tactile perception calculation model, analyzes the related properties to realize synaesthesia and transfers the actual perception to the brain, which is a very challenging technology.

Tradewind Think Tank: What are the similarities and differences between non-invasive and invasive technical principles? And what are their advantages and disadvantages?

Li Yuanqing: At present, there are two kinds of BCI

technologies: non-invasive and invasive. Non-invasive is an external connection and it can record and interpret brain information only with a wearable device, which helps to avoid the risk of brain damage caused by craniotomy. But it receives low and inaccurate signal, because the external equipment is far away from the cortex and is separated from the skull, so the EEG signal acquisition is greatly affected by the ambient noise.

In fact, EEG is a typical non-invasive system, which is cost-effective, easy and simple to use. EEG is very sensitive to noise, and its precision control is easily affected.

The invasive implants electrodes into the brain through surgery to obtain high-quality brainwave signals, which more directly intervenes and regulates the brain, making it more convenient for the brain to control external machines, but causes a relatively high risk and cost for surgery. In addition, foreign material invasion may affect brain immunity and wounded tissue, leading to the deterioration or even disappearance of the electrode signal quality, and the incurable wound even occurs inflammation. Almost all of the current commercial BCI systems are non-invasive, and invasive BCI system is mainly used in the research of special patient treatment. However, with the further improvement of implantation technology, the invasive BCI will be widely used in the future.

Tradewind Think Tank: What will be the mainstream exploration direction between the non-invasive and invasive modes? In which areas will they generate applications respectively?

Li Yuanqing: In fact, there is no superior or inferior between non-invasive and invasive BCI technology. They are two completely different models, and both have their own applications. Invasive BCI, which requires an implant surgery, may be harmful to the brain, but they are better at capturing and processing electrical signals and are more direct and effective in neuro modulation. Hence invasive BCI mainly uses to diagnose and treat patients with epilepsy, Alzheimer's, Parkinson and other mental illnesses.

Musk's Neuralink company focuses on research with invasive BCI. Although the research has not yet been formally applied to the human brain, Neuralink has already started animal trials with promising results. In February 2021, Neuralink successfully enabled a monkey to play a video game of table tennis with a brain chip. It was able to record and decode electrical signals from the monkey's brain by inserting more than 2,000 electrodes into the motor cortex on both sides of the brain, an area that coordinates hand and arm movements. When the monkey tries to play the game, the brain waves generated by the neurons in the brain are read by the BCI device, recording which neurons are activated and sends the data to a computer decoder. Using the decoded data, the researchers were able to build a mathematical model around the relationship between the monkey's neural activity patterns and the hand movements that manipulate the joystick, and calibrate the decoder to predict hand movements based on the monkey's brain activity. At first, the monkeys were trained to play with their right hand using a joystick. As their brain awareness was further synchronized with the decoder, the monkeys were able to imagine moving hand to control game commands after disconnecting the joystick.

Intrusive BCI technology has made great progress not only on monkeys, but also in humans. Nature, for example, recently reported an invasive BCI that turned "thoughts" in the human brain into words on a screen, with great speed and accuracy.

The non-invasive BCI is not harmful to the brain, but its intervention on the brain is not as direct and precise, so non-invasive technologies are more likely to be used in areas such as education, entertainment and smart homes.

For example, in education, we can develop many BCI application systems, including developing personalized training programs, training attention, and relieving learning stress.

In terms of entertainment, non-invasive BCI technology can be combined with virtual reality technology, thus, we no longer need additional control equipment. Directly through the mind to control the game character, we can get a more immersive game experience in the game world. For another example, in home automation, the BCI technology combined with the Internet of Things can achieve "remote control" for doors, windows, TVs, air conditioners, and lights through users' minds, to further enhance the sense of technology and convenience of the home.

What Industry Will Be BCI Applied?

Tradewind Think Tank: What will be the main focuses for the next industrial application of BCI? And what problems need to be solved?

Li Yuanqing: Based on the existing commercial applications and future industrialization exploration, the BCI will develop in three directions: state recognition and monitoring, information exchange and control, perception and motor function for rehabilitation and enhancement.

The first direction is state recognition and monitoring. BCI is entering the fields of education, entertainment, and work management. In the field of education, BCI devices can provide real-time assessment of students' attention, help teachers get real-time feedback on teaching effectiveness, and provide reference for improving teaching content. In terms of work management, when special working positions, such as airline pilots, air traffic controllers and long-haul truck drivers, become fatigued, BCI can be applied to monitor the brain data of these special positions in real time, providing an important objective evidence for work management.

In these commercial scenarios, non-invasive BCI solutions will be the mainstream, and relevant products will be portable wearable devices. In the past, the acquisition of brainwave signals had to rely on complex instruments and equipment, requiring the subject to wear a heavy, cable-filled hat. In the future, with breakthroughs in new materials and signal processing algorithms, portable wearable brain-computer devices will become more and more mature.

The second direction is information exchange and control. In this regard, more and more brain-controlled external devices have gradually emerged, such as brain-controlled robotic arms, prostheses, and remote controls. The core is to achieve direct brain control of machinery and equipment through the BCI. The Harvard University research team invented a smart myoelectric prosthesis based on BCI technology-BrainRobotics. The prosthesis uses artificial intelligence algorithms to process brain waves and Electromyography (EMG) signals, and uses surface EMG sensors to detect the muscle activity of the residual limbs of disabled patients, so as to train patients to actively contract muscles to allow the prosthesis to perform a variety of operations. Based on this technical principle, patients can achieve extremely difficult behaviors such as writing calligraphy, playing the piano and rock climbing.

The latest global research result in this direction is that in 2020, Johns Hopkins University in the U.S announced the first realization of simultaneous control of two robotic arms. Controlling two robotic arms simultaneously with the help of brain-computer interface is a more difficult challenge, because it is not a simple left-arm algorithm+right-arm algorithm, but a general cognition and resolution of the target task, and then assigned to both arms to complete together. It needs to incorporate more overall planning, correlation and coordination into the algorithm model. It is much more difficult to control two robotic arms with BCI than to control a single arm.

The third direction is the rehabilitation and enhancement of perceptual and motor functions. At the opening ceremony of the 2014 World Cup in Brazil, a paralyzed young man managed to kick the first goal of the tournament with the help of BCI and a mechanical exoskeleton.

Previously, this young man had a high degree of paralysis, with only 4 healthy spines. With the help of BCI technology, the movement consciousness is transmitted to the mechanical exoskeleton, which drives the movement of the limbs. After about 10 months of training, it successfully restored the perception and movement control functions of 11 vertebrae. This can be said to have been repaired from "completely paralyzed" to "partially paralyzed."

In the future, the "BCI+mechanical exoskeleton" combined rehabilitation program and related products will have huge application value and market potential, which can bring hope to tens of millions of patients worldwide.

However, BCI faces some problems. One is how to solve the problem of user acceptance. Whether it is invasive or non-invasive brain-computer interface, it is not very easy for the average user to accept. Because BCI technology is strongly related to the brain, its safety has not yet been recognized by most people, or that it has not yet formed an official or industry standard, so most of the applications now are for very specific patients. This is a key point where brain-computer interface technology must be enhanced.

Another is the balance between performance and cost in the productization process. It requires that the product must effectively solve user needs because users pay new technical products for its intuitive effects. And it also requires to balance the cost between company's technical equipment development and production. If making a profit is short-lived, industrialization cannot go further.

Tradewind Think Tank: Medical is one of the important application areas of BCI. Could you please tell us what solutions and application achievements have been made in the medical field?

Li Yuanqing: Take mind texting as an example. A foreign research team has developed an invasive BCI that successfully allowed a 65-year-old paralyzed patient to write words with minds, and then display the typed version of these words on the computer screen in real time, with extremely high accuracy. So how is it done? The working principle is by implanting two tiny electrode arrays in the patient's brain to collect and receive the brain's control of handwriting, that is, imagining handwriting in the brain, and then build an algorithm model, to decode and converse the brain information, displaying the image on the screen.

In fact, the underlying algorithmic logic of the technology is based on machine learning. Mindful handwriting requires a lot of training to further improve accuracy. Especially for very similar characters like "2" and "Z," handwriting is difficult to distinguish, which requires the algorithm model to carry out a lot of text training and comparison. For example, the front and back of "2" are numbers, and the front and back of "Z" are letters.

Then take the first invasive BCI adjuvant therapy in China as an example. A 72-year-old patient suffered a car accident a few years ago, leaving her limbs completely paralyzed, with only her brain intact and undamaged. The Second Affiliated Hospital of Zhejiang University Medical College successfully helped the patient to accurately control the external robotic arm and manipulator by using EEG signals through an invasive BCI, and to restore some mobility.

This is not an easy process to implement. Although the EEG signals from the patient's brain can be recorded in real time through an invasive BCI, controlling the robotic arm is not easy because it requires the instructions from the brain signals to be fully adjusted and coordinated with the movements of the robotic arm. For example, patients imagine how to control the swing and strength of the robotic arm when it moves, and how to accurately make the robotic arm reach the desired position, which requires repeated machine learning and manual training.

The research team of Zhejiang University used a self-developed artificial intelligence algorithm to decode the electrical brain signals sent by the patient and then direct the robotic arm to move more naturally and smoothly. It also took the elderly patient four months of interactive training to achieve smooth mind control. The patient is now able to use a mind-controlled robotic arm to pick up a Coke from a table, insert a straw, and deliver it to his mouth with precision.

In addition, the South China Brain-computer Interface Technology Co., Ltd (IHNNK) also has made some achievements and applications in medical-related fields, such as consciousness detection. For patients with severe brain injury, we use visual stimulation or auditory stimulation, and then use BCI to detect the response of the patients, so as to assess whether the patients have facial recognition, number recognition, emotion and other functions, and then judge the degree of consciousness impairment of the patients. Another application is motor function assistance. For those patients who lack mobility, we have developed a set of intelligent ward system with brain-machine AI technology as the core, which can effectively ensure that patients send call instructions to nurses through head-mounted electroencephalogram interactive equipment, and can also realize simple control of lighting, air conditioning, nursing bed and other equipment.

The application of BCI technology in the medical field is of great value. In addition to the rehabilitation effects for patients with physical and motor dysfunction, it has great potential for the detection, auxiliary diagnosis and rehabilitation prediction of patients with mental disorders. Technical research in these directions has already yielded some results, and it is believed that more practical applications will be implemented in the near future.

What Is the Imagination of BCI?

Tradewind Think Tank: What problems still exist in the development of BCI technology at this stage? How will brain-computer intelligence break through in the future?

Li Yuanqing: BCI is an emerging interdisciplinary field. Although some important breakthroughs have been made at this stage, such as assisting the diagnosis and treatment of some human diseases, there is still a lot of research that remains in the laboratory stage, and there is still a considerable distance from the large-scale universal application.

The further application development of BCI technology still needs to break through the following three bottlenecks. First of all, the most important and difficult point is that there has not been a substantial breakthrough in the research of "brain science," and the human understanding of the brain is still very superficial. Even though everyone has realized that the brain can issue instructions to control behaviors, but do not understand the process of instruction formation. Thus, it is impossible to read instructions effectively. For example, when we use a mouse and keyboard, we need arms and ten fingers to cooperate with each other to operate with reasonable time, position and strength, and finally achieve the results we want. In terms of the actual movements, many people may find it particularly easy to type, requiring only the movement of their fingers. But if we look at the logic of its operation, we will find that there are many neurons in the brain that participate in this process and they are distributed in different brain regions. Their signals are very complex, and it is extremely difficult to collect a complete signal, and decoding is even more difficult. With the current technology, we can at most capture some of the neuronal signals in specific brain regions,, and then combine them with the latest data analysis methods in the hope of deciphering the brain's signal commands from small ones, so the accuracy cannot be 100%.

Secondly, there are still some challenges in the development of BCI hardware equipment. Since we want to collect brain signals, the BCI requires specific hardware equipment. On the one hand, hardware access that requires a BCI does not harm the brain as much as possible; on the other hand, hardware devices are required to be as small and convenient as possible, functionally stable, and cost-effective. Obviously, the current BCI equipment materials can only meet some of the conditions, and there is still much room for improvement and research and development.

Finally, the current application range of BCI is still relatively limited to the medical field, and there is less coverage in the ordinary life field. Although some researches on user applications have been conducted, such as manipulating a mouse through a BCI and playing games, there are still many restrictions and have not fully entered daily life. How to develop some applications that can arouse public interest requires continuous exploration by R&D institutions and technology companies.

In general, BCI, as an emerging field, has unlimited imagination, and realizing the connection between human brain and computer (artificial intelligence) is only the first step.

Tradewind Think Tank: With the continuous development and breakthrough of BCI technology, concepts such as consciousness storage, consciousness reading, and human-computer symbiosis seem to become a reality. How should we understand and view these assumptions?

Li Yuanqing: In fact, what is often referred to as "mind reading" or "telepathy" has already been achieved to a certain extent by current BCI technologies. "Mind texting" via BCI is essentially a form of "mind reading," in which we are able to express whatever is going on in our minds.

"Mind reading" is actually the future of technology itself and the development of such technology will bring transformative positive value. For example, we could use this technology to identify or alert in advance who is suffering from schizophrenia, dyslexia, autism or Alzheimer's disease. We could also look at patterns of brain activity to determine if a person is at risk of suicide, so we could intervene before it's too late. For another example, through BCI technology, we can better understand human emotion perception system such as fear, anxiety, attraction, and pleasure, so that we can create products and services more in line with users' needs and improve people's happiness in an all-round way.

At the same time, it should be noted that behind the positive value, there are usually negative effects, and a new technology will often face moral and ethical problems in the process of innovation. We know that while Big Data brings economic value, it also brings issues of data privacy and security. If BCI technology is further developed, it will undoubtedly face similar problems, which will require government departments, scientific research institutions, industry and other sectors of society to jointly formulate relevant regulations and standards. Of course, in terms of the current development stage, BCI is still in the early stage of technological exploration, and the focus is still on technological innovation, iteration, and maturity.

As for the "consciousness input and storage" that everyone talks about, it is just an imagination or a concept, which is far from being realized. However, from a macro point of view, the integration of human intelligence and artificial intelligence is indeed the long-term goal of many research institutions.

(According to the online data, the Tradewind Think Tank)

Some Domestic Brain Research Institutions
(in no particular order)

Institution Name	Time of Stablishment	Introduction
Institute of Neuroscience, Shanghai Institutes for Biological Sciences, Chinese Academy of Sciences	1999	Dedicated to various fields of basic neuroscience research, including molecular, cellular and developmental neurobiology, system and computational neuroscience, and cognitive and behavioral neuroscience.
Beijing Brain Science and Brain-Like Research Center	2018	Focusing on five aspects: common technology platform and resource library construction, major diseases related to cognitive impairment, brain-like computing and brain-computer intelligence, brain development of children and adolescents, and analysis of the principles of brain cognition.
Shanghai Brain Science and Brain-Inspired Research Center	2018	It will mainly build a technical support and R&D sharing platform, undertake major scientific research tasks at the Shanghai level, as well as major projects of the China Brain Project and the international large-scale scientific plan of the "Whole Brain Mesoscopic Neural Connection Atlas" Research tasks.
Fudan University Brain Science Frontier Science Center	2018	The first frontier science center of the National "Everest Project," focusing on brain-like intelligent technology research and brain-like chip intelligent system development, with the characteristics of close integration of multidisciplinary and clinical research.

Institution Name	Time of Stablishment	Introduction
The Frontier Science Center for Brain and Brain-Machine Fusion, Zhejiang University.	2018	We explored and promoted the convergence and fusion of brain science and artificial intelligence. The key research areas are: the unique molecular and cellular biological mechanisms of the nervous system, the mechanism research and intervention measures of brain diseases, and the brain-machine interface and hybrid intelligence, brain-like computing and neural computing, etc.
IDG McGovern Institute for Brain Science (jointly established)	2011	Peking University IDG/McGovern Institute for Brain Research.
	2011	The IDG/McGovern Institute for Brain Science of Beijing Normal University was jointly established.
	2013	Tsinghua University IDG/McGovern Institute for Brain Science was jointly established.
	2014	Shenzhen Institute of Advanced Technology, Chinese Academy of Sciences and MIT McGovern Institute for Brain Science jointly established the Institute of Brain Cognitive Science and Brain Disease.
Innovation Center of Excellence Brain Science and Intelligent Technology, Chinese Academy of Sciences	2014	The center's research work mainly includes five areas: the circuit basis of brain cognitive function, brain disease mechanism and diagnostic intervention, new brain research technology, brain-like models and intelligent information processing, and brain-like devices and systems.

Institution Name	Time of Stablishment	Introduction
Sichuan Institute of Brain Science and Brain-inspired Intelligence	2018	Focusing on brain-inspired intelligence and neural engineering, focusing on exploring brain mechanisms, diagnosing and treating brain diseases, and imitating brain intelligence, and building a cloud-brain big data platform and a brain-inspired intelligence research platform. To carry out research on brain science and brain-like intelligence technology.
Chongqing Brain Science Collaborative Innovation Center	2015	The Chongqing Brain Science Collaborative Innovation Center was led by the Third Military Medical University, and was jointly constructed by Southwest University, Chongqing University, Chongqing Medical University, Chongqing Normal University and Chongqing College of Arts and Sciences. The three basic directions of "cognition, brain therapy, and brain development".
Guangdong-Hong Kong-Macao Greater Bay Area Brain Science and Brain-inspired Research Center	2018	Relying on Southern Medical University and Guangdong Province to promote and converge the Guangdong-Hong Kong-Macao Greater Bay Area brain science and brain-inspired research related resources to jointly build a research and development institution. The center is dedicated to brain science and basic and applied research on brain-like intelligence technology.
Research Institute Of Shenzhen-Hong Kong Brain Science Innovation	2019	Closely focused on the core issue of brain cognition and the neural mechanism of brain diseases, and focused on the "neural basis of cognition," "the mechanism of brain diseases," "diagnosis and treatment strategies for brain diseases" and "new brain science research." Research on four key areas of "Technical Methods."

Institution Name	Time of Stablishment	Introduction
Heilongjiang Provincial Brain Science and Brain-like Intelligence Research Center	2019	To carry out research on brain science, brain-like intelligence, and BCI, including research on pathological mechanisms of mental illness, cognitive rules and neural mechanisms, brain-controlled exoskeleton and external muscular system, smart medical care and brain health platform research and development, etc.
Henan Provincial Key Laboratory of Brain Science and BCI Technology	2016	Comprehensive utilization of innovative advanced technologies in neuroscience, information science, control science, materials science, microelectronics technology, etc., from neurons, neural circuits and brain function networks, etc. Hierarchical acquisition and analysis of brain information, breakthrough key technologies of BCI.
Shandong University Institute of Brain and Brain-inspired Science	2015	Mainly carried out research on brain structure and function, early diagnosis and intervention of brain diseases, brain dynamic technology and brain-inspired intelligence, etc.
Wuhan Brain Science Center	2019	Aiming at the international frontiers of brain science, taking the lead in undertaking the key tasks of the "China Brain Project," and focusing on the three main research areas of "human brain working principle," "functional visualization and brain-like computing" and "brain disease diagnosis and treatment." Carry out brain science and brain-like research.
Tianjin Brain Science Center	2019	Mainly explore the neural mechanisms of advanced cognitive functions of the brain through physiological, pathological, and behavioral methods, and carry out a new generation of brain-computer interaction and brain-like intelligence research.

(According to the online data, the Tradewind Think Tank)

Company name	Country	Technical Route	It's Not Played
Zhejiang University Xitou Brain Machine Intelligent Technology	China	Invasive/	Become a brain-computer key technology and product prototype research and development base, build a "brain-computer-brain" two-way interactive and open national platform, and practice the deep integration of production and research: adhere to the demand-oriented and boost the development of brain-computer intelligence related industries.
South China Brain Control Intelligent Technology	China	Non-invasive	The R&D direction is mainly focused on the education and medical fields.
Kedou Brain Machine Technology	China	Non-invasive	Development of software and hardware related to BCI, products are mainly used in medical and education fields.
Burun Technology	China	Non-invasive	EEG acquisition, EEG analysis, EEG pattern recognition, BCI control and application.
Bo Rui Kang Technology	China	Non-invasive	Committed to providing professional and complete solutions for neuroscience innovation research and clinical neurological disease diagnosis, treatment and rehabilitation research.
Brainland Technology	China	Non-invasive	Focus on cutting-edge brain science technology applications such as brain science, brain health screening, EEG algorithms, EEG data open platform, and brain-like decision-making computing.

Company name	Country	Technical Route	It's Not Played
NeuraLink	United States	Non-invasive	Develop high-bandwidth and safe and reliable BCI technology.
Brainco	United States	Invasive	Committed to the development of non-invasive BCI technology, research and development of wearable devices based on brain waves, to achieve the goals of attention training and functional recovery of patients with hemiplegia.
Kernel	United States	Non-invasive	Research on human intelligence and devote to research and development of neuroprosthetic technology.
NeuroSky	United States	Invasive	Development of wearable devices controlled by brain waves.
Emotiv	United States	Non-invasive	Develop simple and easy-to-use mobile wearable devices that can monitor the cognitive and emotional state of users.
Mindmaze	Switzerland	Non-invasive	A platform that combines VR, brain imaging, computer graphics and neuroscience.
InterXon	Canada	Non-invasive	Develop a brain wave monitoring headband that can help users meditate and relax through real-time audio feedback.

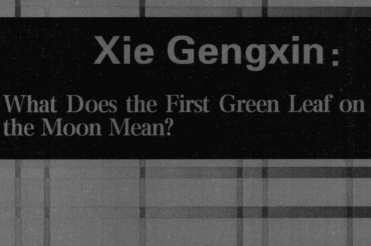

Xie Gengxin:

What Does the First Green Leaf on the Moon Mean?

What is the main purpose of deep space exploration? What is its impact and significance on our national production and life? What is the connection between deep space exploration, data collection and application, and how does data promote scientific research in this area? How to use the new generation of information technology such as big data intelligence in the field of deep space exploration, and how to pull the development of big data intelligence technology in deep space exploration?

Demystifying the Far Side of the Moon

Tradewind Think Tank: Chang'e-4 lunar probe is the first probe in human history to land and explore the far side of the moon. What important scientific research achievements has it made so far?

Xie Gengxin: Chang'e-4 lunar probe landed on the far side of the Moon and scored a series of scientific and engineering achievements. Chang'e-4 lunar probe has been working since January 3, 2019, and has collected environmental data on the moon, including temperature and solar wind, as well as data on the topography and structure of the lunar surface. The large amount of first-hand information has laid a foundation for building lunar research stations and lunar bases for manned lunar exploration, and has also served as a good scientific guide for scientists to study the origin of the Moon and the development and utilization of lunar resources.

Tradewind Think Tank: What is the difference between Chang'e-4 lunar probe exploration on the far side of the Moon and the previous explorations on the near side of the Moon?

Xie Gengxin: In fact, this is not the first observation of the far side of the moon by human beings, because many flying vehicles also fly around the Moon, and some observations are also being made. But this is the first time a spacecraft has landed on the far side of the moon for in-situ detection.

The far side of the Moon is actually similar to the near side. It's just that humans used to detect the moon through observations, thinking that the far side was mysterious, and translated it to dark side before for it was invisible from the Earth. The Moon rotates around the Earth at the same time as it rotates. Observations from the Earth can only see the far side not the near because of tidal locking.

With the development of technology, we set up a relay satellite into the lunar orbit, Queqiao (Magpie Bridge), to transmit the image data and signals from the far side of the Moon back to the Earth. We found no fundamental difference between the far side of the Moon and the near side. So now we call the far side, which means the side slightly farther from the Earth than the near side.

The First Green Leaf on the Moon

Tradewind Think Tank: On Chang'e-4 lunar probe, the biological science demonstration load project led by you has grown the first green leaf on the moon. From the perspective of deep space exploration and biological ecology, what significance does it have?

Xie Gengxin: In this biological experiment, we constructed a closed ecological microenvironment. Besides the availability of air with atmospheric pressure, suitable temperature and humidity, other conditions are to strive for the true conditions of the Moon's surface. For example, the one-sixth low gravity field, the strong sunlight of the Sun, and the lack of any radiation protection measures. Such a special space environment cannot be fully simulated on the Earth.

So, in the very special environment on the Moon, whether life can adapt to the environment, how to adapt to the environment, whether it can grow and how to grow, all have great uncertainty, so it is very worthwhile to verify. The first significance is a pioneering experiment. Hence, *Nature* described our experiment as a "pioneering experiment".

As Professor Mike, chief scientist of biology at NASA, said, "If we take a seed to the Moon, if it doesn't die in 14 days, then it means the seed can grow on the Moon." Fortunately, in our experiment, the seed did not die, but grew a green leaf. Therefore, organisms can adapt and grow in this special environment by constructing an environment suitable for survival.

In the experiment, the air, soil and water on the Earth were sealed with tanks. But there were still several conditions that were not deliberately created. For example, we let the Sun on the Moon shine directly into the tank for photosynthesis, without considering the light intensity and radiation protection on the tank material, maintaining the real light conditions. In the next step, we will consider making more use of the original conditions on the Moon, such as using lunar soil, or even making water on the Moon. Now everyone is following up on the technology of in-situ water generation on the Moon, and it has been accumulated and feasible to some extent. The originality of this experiment has been fully affirmed internationally and science evaluated our results as "in a first for humankind."

The second significance is display. The experiment was designed with display in mind. Photographs of the seed's growth on the moon must be taken and sent back to the Earth. In addition to the value of scientific research, it can allow young people to understand how to create such a special environment on the Moon and how organisms grow in it so as to stimulate their enthusiasm for scientific exploration. Similarly, engineers and scientists have paid great efforts to complete this experiment — a leaf grown on the Moon, which can awaken the public's awareness of environmental protection and let everyone know the importance of environmental protection on the Earth. Our team has carried out more than 120 popular science lectures in Chinese universities, primary and secondary schools, and was invited to the UK, Spain, France and Sweden to carry out exchanges, disseminating scientific knowledge and publicizing the achievements of China's lunar exploration program, which has been well received by everyone.

The third significance is the realization of the project. It took a very small cost to do such an experiment. The total mass of the experimental device is 2.608 kilograms, including two cameras, control panels, insulation materials, data cables, aluminum alloy cans, 18 grams of water and soil. Scientists have constructed a controllable ecological environment system suitable for the growth of life on the Moon in a very simple and feasible way. If it is enlarged, its next step may become a farm with a variety of vegetables, or a garden with plants. In fact, this is the first green plant that humans have grown on extraterrestrial planet, which is the project significance. This can provide a basis for exploration in the future for manned Moon landings and even survival and travel on the Moon. Therefore, the next step is to summarize the experience, further study and design on this basis.

Tradewind Think Tank: What ecosystem considerations did you and your team make when designing the project?

Xie Gengxin: The experiment was designed with our goals in mind. With future survival on extraterrestrial planets in consideration, we chose species based on human necessities such as potatoes, which are suitable as a staple food in space; cotton, represents the material of clothing; rape, stands for cooking oil; fruit flies, represent animals and are also widely used in experiments on genetics and evolution; Arabidopsis thaliana, a light-loving plant, or the fruit fly of plants, is a good model organism for research, especially for genetic experiments in plants; and yeasts, which represent microbes. We not only allow animals and plants to survive and grow, but also want them to form a controlled closed ecosystem, where plants use carbon dioxide for photosynthesis to produce oxygen, animals use oxygen for respiration to produce carbon dioxide, and the decomposition of microorganisms to form micro-ecological cycles.

The animals were also chosen for display and publicity as the experiment was released globally. The preferred animal is a tortoise as it has a shell to protect itself from injury during liftoff. The tortoise is also very popular, and it is an auspicious animal that represents longevity in Chinese culture. In addition, Chang'e 3 and Chang'e-4 carry the Yutu-1 and Yutu-2 lunar rovers respectively, so the tortoise is a reference to the race between hare and tortoise. Unfortunately, Chang'e-4 has very limited resources for this experiment. Even if we let the tortoise enter the dormant, it can only stay in the tank for more than 20 days before all the oxygen is used up.

The space given by the Chang'e-4 payload project is limited to just 200 millimeters by 180 millimeters, including the tank and the outer insulation, so the biological space inside is only 0.82 liters. Some people asked us why we don't carry more pure oxygen. This is because there are only two control channels on Chang'e-4, one for water discharge and one for taking pictures, and there's no extra control channels for gas release, and no extra mass for us to add other parts. Therefore, we can only fill a 0.82 liters space with a standard atmospheric pressure air, which contains 22% oxygen.

Then we considered silkworm for animal experiments. Silkworms are also very auspicious, representing Chinese silk culture and the Silk Road. Moreover, silkworms wrap themselves in cocoons to protect themselves during launch. Silkworms, however, are very delicate and may freeze to death in a low temperature environment facing away from the sun. So, we tried to install insulation with batteries as an energy source. However, a third-party assessment believed that the battery may cause an explosion, so this plan was eventually abandoned.

We put a lot of thoughts into designing the experiment. First, it must be scientifically meaningful and feasible. Second, there must have practical needs, such as food and clothing. Third, there must have a value for the inheritance and dissemination of Chinese culture, to promote China's achievements and culture to young people, the general public, and the international community.

As you can see, the cotton seeds we planted on the Moon

sprouted on January 4, 2019, and the leaves were still green until the shutdown on January 12 or even the fifth month. This shows that photosynthesis is still working, and carbon dioxide is being generated. It also shows that our fruit fly animals are also living in it. Unfortunately, because of very limited resources, our point-and-shoot cameras can only take pictures of cotton buds, not other species.

Only a Nation Looking at the Stars Has a Future

Tradewind Think Tank: What is the main purpose of deep space exploration? What impacts and significance do it have on our national production and life?

Xie Gengxin: In fact, no matter how much technology develops, there are two things we must face: One is how people actually came to be; the other is how the world we live in came to be and what it is all about.

In ancient times, people didn't understand lightning and thunder. They didn't know that it was originally caused by clouds and ionization, instead, they thought it was the function of gods. In the past, we didn't know that there were wireless microwaves in our space. Later, when we discovered wireless microwaves, we used it to realize mobile phone wireless communication.

The whole universe is very mysterious and the exploration is endless. Deep space exploration is generated by the desire and curiosity of human beings to explore the unknown. It expands our understanding of the space dimension of the universe step by step to a deeper and farther place. That's what's driving our imaginations to go to the Moon, travel into space, and orbit the Earth in a synchronous orbit. Therefore, rockets, satellites and space stations came into being. This process has naturally led to the development of science and technology, economy, and society.

Deep space exploration is not only driving the distance and dimension of the universe deeper and farther, but also the great progress in technology and thinking. When looking at the earth from space and thinking about life above the universe, our outlook on life and values will change. As the German philosopher Hegel said, "A nation there is a group of people looking at the stars, they have hope." If a nation does not have such people, this nation is destined to have no future. This is the discussion of deep space exploration from philosophy. Regardless of feelings, pursuits, or dreams, it is precisely these things that promote social progress. It's not that how much money you make will promote social progress.

This is also why countries all over the world attach great importance to deep space exploration, as it is an important means to promote social and human progress. That's why Armstrong said, "That's One Small Step for a Man — One Giant Leap for Mankind."

Tradewind Think Tank: What is the relationship between

deep space exploration, data collection and application, and how can data contribute to scientific research in this area?

Xie Gengxin: Only about 10 astronauts in China have been in space on spacecraft, and neither scientists nor engineers have such experience. Then, our research can only be based on a large amount of scientific data on deep space exploration. For example, we have never been to the Moon, so how to design the carrier tank of this biological science demonstration load project, what its shape should be, how big its size should be, and how to control the temperature of the material, these designs are completely based on a lot of data accumulated before. Therefore, the first important result of deep space exploration is data.

Whoever gets the data will have the discourse power in this field. For example, we got a lot of data from the first landing on the far side of the moon. Also, the Chang'e-5 probe retrieved lunar soil, including some data about the lunar environment. Only after getting these massive, first-hand lunar exploration data, can scientifically research on the moon be carried out. Otherwise, we can only use foreign data and cannot achieve a real international lead. The data collection from these deep space probes, including environmental, astronomical, meteorological, and orbital aspects, are very valuable.

An important guarantee for deep space exploration is the Deep Space Monitoring and Control Network. It has deployed data receiving sites and control sites throughout the country and even around the world, which are used to connect with space probes to obtain, release, and transmit data. Also, through this network, we can control space probes through digital signals. Throughout the process, data plays the first key role. If our data is not right and the algorithm model is not good, the mission will fail. So for the data of deep space exploration, on the one hand, we have to manage it in an orderly way; on the other hand, we have to release it in an orderly way, and some data can be shared with the whole world.

Tradewind Think Tank: On a global scale, what changes have occurred in today's deep space exploration compared to the past in terms of technology, mode and pattern?

Xie Gengxin: For scientific and technological exploration, the space environment has incomparable value. For example, environmental conditions such as radiation in space, and high and low temperatures can make genetic modification or improvement, which is unmatched by radiation or ultraviolet rays on the Earth. Many biological companies send some seeds into space, then cultivate and optimize them after returning to the Earth to obtain more high-quality genetically-modified varieties. For example, many cutting-edge researches in science and technology require vacuum or ultra-low temperature conditions. It is very difficult and costly to achieve these conditions on the Earth, but it can have natural environmental conditions in space. Therefore, this is one of the major reasons why countries all over the world attach great importance to deep space exploration. Our country, with the development of science and technology and economy, also officially launched the lunar exploration project as early as 2004.

In terms of technology, today's deep space exploration has undergone great changes compared with the 1960s. The first is that communication control is more convenient and simpler and the equipment channel is more advanced. The second is a breakthrough in material technology, which makes the material itself lighter, and its hardness and radiation resistance have been greatly improved. The third is a significant increase in computing power. In the past, a computer was as big as a house, but now it only needs a mobile phone to exceed the previous computing power.

In terms of the model, today's deep space exploration and lunar probe are not like the space race between the U.S. and the Soviet Union during the Cold War for national interests and political contests, but more closely integrated with technological progress and industrial development. Each participating country is thinking about how this project can promote the progress of science and technology; how new technologies, new techniques, new materials and new methods can be verified and tested in space; how to get more commercial and civilian companies involved; how to industrialize and commercialize the achievements of deep space exploration as soon as possible.

The U.S. has handed over many space exploration projects to commercial companies and has clarified the business and results of commercial companies. In this way, the results of these space projects can be applied to all aspects of the economy and society more quickly. Therefore, the number of launches by commercial companies is increasing, and the frequency of exploration is getting higher and higher.

In terms of pattern, international cooperation has been strengthened. Previously, only the U.S and the Soviet Union were capable of deep space exploration. But now, many countries are involved. Additionally, the cooperation between countries has increased. Not long ago, China and Russia planned to build a lunar research station. In this way, the International Space Station becomes an international stage, not a solo act by one or two countries. The global pattern has changed, forming a new pattern of global cooperation.

Tradewind Think Tank: How does the new generation of information technology such as big data intelligence apply to the field of deep space exploration, and how does deep space exploration drive the development of big data intelligence technology?

Xie Gengxine: Big data and artificial intelligence have a huge boost to deep space exploration. For example, intelligence makes exploration activities easier to carry out and detection data easier to obtain. For example, an instrument on the moon can not only automatically collect and transmit data back to the Earth, but also can be in-situ in place to automatically analyze the composition of the data and results. A lot of data transmitted from the Moon has been automatically analyzed and formed into results, which is conducive to researchers on the ground to study, and the data is more accurate and real-time.

Currently, in lunar exploration and deep space exploration, more intelligence is used for unmanned exploration or unmanned and manned collaborative exploration. The Mars exploration of China's "Tianwen-1" and the U.S "Perseverance" forms a data packet through many automated and intelligent in-situ exploration, and then transmits it back to the Earth for research. The same is true for the lunar exploration project. In fact, every deep space probe is a highly intelligent robot that replaces humans in high-risk space exploration work.

In turn, deep space exploration provides a unique application scenario for the development of big data intelligence, which greatly pulls the improvement of technology level and the verification of new technologies.

On the one hand, deep space exploration has obvious application requirements and clear mission objectives, so it can accurately influence the improvement and realization of technology to meet the needs and achieve the goals. For example, autonomous navigation and unmanned driving are not yet available for practical use on the Earth due to legal and ethical issues. But on the Moon and Mars, that won't be the case because the probes are unmanned.

On the other hand, deep space exploration can also play a role in technical verification, verifying the feasibility and scientificity of an emerging technology in an extremely special environment. It's like if a person can survive in an extreme environment, then he can adapt more easily and survive better in an ordinary environment. Deep space exploration has brought many extreme challenges to technology, just like a technological extreme sports arena, which can carry out some preliminary verification of many cutting-edge technologies.

Tradewind Think Tank: Would you please tell us about the planning of China's lunar exploration project and the evolution of the series of Chang'e lunar probes?

Xie Gengxin: What Does the First Green Leaf on the Moon Mean?

Xie Gengxin: Since 2004, China has set up a project to explore the moon, with a long-term plan of three steps: exploration, landing and station. Exploration is the ongoing lunar exploration project, that is, to explore the Moon, and to find out the overall situation of the Moon. Landing is to have astronauts on the Moon, boarded to step a footprint. We haven't realized this step yet, and our Tiangong space station is for that. After landing on the Moon, our next step is to establish a base for astronauts to stay on the Moon.

Now we are in the first step of "exploration", and the lunar exploration project is divided into three small steps: "orbiting, landing, sample returning." The first task is to "orbit" and revolve around the moon, which is carried out by Chang'e-1 and Chang'e-2. Previously, through international literature, reports and our own astronomical observations, we knew the topography, environmental geography and astronomical space of the moon data and information, but this is not our own first-hand data after all. Therefore, we first launched the Chang'e-1 to just revolve around the moon. It just took a look and used optical, microwave and radar methods to scan the moon. Chang'e-1 and Chang'e-2 took the lunar digital evaluation model pictures of the whole month.

The Chang'e-1 to Chang'e-6 are paired and backed up each other. That is to say, Chang'e-1 and Chang'e-2 have the same goal and design, and its development, production and delivery are all carried out at the same time with the same status. The launch of Chang'e-1 was successful, if not Chang'e-2 will immediately follow.

What if Chang'e-1 succeeds but Chang'e-2 has been developed? In addition to getting a full moon map, Chang'e-2 will have to lower its orbit. Chang'e-1 was designed to look at the Moon in a large circle, and because of the success of Chang'e-1, Chang'e-2 can carry out more dangerous missions. While Chang'e-1 was in an orbit thousands of kilometers from the Moon, Chang'e-2 lowers its orbit dramatically and circles the Moon in an elliptical orbit of 100 kilometers by 15 kilometers. In this way, we can see more clearly and get a 7-meter resolution image of the Moon.

More importantly, Chang'e-2 also paved the way for Chang'e-3 and Chang'e-4 by detecting the landing points of 3 and 4, especially the 1.5-meter-resolution Moon map of the front of Hongwan Bay. Therefore, Chang'e-1 and Chang'e-2 obtained a large amount of first-hand lunar data, laying a solid foundation for Chang'e-3 and Chang'e-4 to land on the Moon.

The missions of Chang'e-3 and Chang'e-4 are to "land" on the Moon. They are also backups of each other. Chang'e-3 succeeded. Though the Yutu was not well controlled. It only ran for a few tens of meters before stop, but it also got the data and sent back photos. This also means that China has become the third country in the world to achieve a soft landing on the Moon.

Chang'e-3 succeeded, and then Chang'e-4 didn't need to land at Hongwan. Therefore, some adjustments were made to the target of Chang'e-4 to increase the difficulty of the task and to land on the far side of the Moon. The biological science demonstration load project, which was also added after the success of Chang'e-3, was to try to conduct biological experiments on the far side of the Moon. The significance of Chang'e-3 and Chang'e-4 is that we can get the first-hand data of the in-situ data on the Moon, rather than just telemetry.

The mission of Chang'e-5 and Chang'e-6 is "sample returning". "Sample returning" is not only the "returning" of the probes, but also to bring back the samples of the Moon's soil and rocks. Chang'e-5 collected 1,731 grams of Moon soil. We were expecting to get two kilograms back, but when we drilled on the surface of the Moon, we didn't drill very deep, and there was too little soil in the rock. For our research, the deeper the lunar soil is drilled, the greater the research value is, because we have obtained the soil data on the lunar surface through remote sensing and telemetry. The Soviet Union got 321 grams of lunar soil in three missions while the U.S. got 381 kilograms.

As a backup to Chang'e-5, Chang'e-6 will be tasked with sampling as well as a broader exploration of other parts of the Moon. This is because of the need to get as much more comprehensive lunar science data as possible. Chang'e-4, in particular, was able to go to the far side of the Moon and get the first-hand data back, which is our contribution.

Three Phases of the Lunar Exploration Project

Phase Number	Mission	Probe	Introduction
Phase I	Circling	Chang'e-1	Chang'e 1 satellite enters the Earth-Moon transfer orbit through the Earth's phase adjustment orbit, and after achieving lunar capture, carries out lunar exploration in a 200-km circular orbit.
Phase 2	Landing	Chang'e-2, Chang'e-3, Chang'e-4	Phase 2 aims to achieve lunar soft landing and lunar surface inspection survey, etc.
Phase 3	Returning	Chang'e-5, Chang'e-6	Phase 3 project will realize the automatic lunar surface sampling and return, and carry out lunar sample ground analysis research.

Tradewind Think Tank: What positive impacts and promotion will the development of deep-space exploration related industries have on countries and regions?

Xie Gengxin: The industrialization of deep space exploration is most widely involved in the field of aerospace. Aerospace industrialization is now an industry that China is planning and focusing on. For example, China is preparing to introduce the space law. A core part of the law is how to promote orderly competition in the aerospace field and how to promote the commercialization of spaceflight and the industrialization of spaceflight technology.

In the execution of space missions and projects, the country's and even the world's top innovative resources have been mobilized. Therefore, the first positive impact is that experts, engineers, and managers in the spaceflight field have been fully trained. This is an important training ground for us to cultivate talents and the role of this is immeasurable. Why is spaceflight so important now? Because spaceflight can allow some disciplines and talents to develop.

The second positive impact is technology accumulation and transformation. The field of aerospace and deep space exploration is the most technologically advanced. The latest and greatest technologies are used in order to get a probe from the Earth to the Moon at the lowest cost and the lightest mass, or to transport a huge object to the space station. There's a lot of technology that accumulates in the process.

The scientists and engineers in spaceflight field are all striving to achieve scientific research goals. For example, if the space station is to operate safely by 2022, everyone has to do everything possible to control it, maintain it, and ensure its safety, but rarely consider how to transform this technology. We believe that this project that brings together national and even global scientific and technological resources has a lot of potential, and it is worth exploring. If these commercial values are unearthed, they can lead to the rapid development of some new industries, thereby benefiting mankind.

Big data and artificial intelligence are good examples. As astronauts are inconvenient to work in the space station and cannot do many things, the entire management and operation of the space station is fully automated and intelligent. The process of designing and manufacturing the space station focuses on a large number of intelligent scientific and technological innovation resources, which are worthy of our commercial tailor-made development.

The third positive impact is to promote the development of industry. Spaceflight engineering is a training ground, not only for state-owned enterprises and national teams, but also for a wide range of commercial and civil technologies and products. So now China also encourages some private enterprises to participate in such major projects. On the one hand, it can cultivate the new technology of private enterprises and the spirit of private entrepreneurs to take responsibility. On the other hand, private enterprises can in turn make the industrial ecology healthier and develop faster.

At present, China's space infrastructure has initially taken shape. How to further develop and utilize deep space exploration, aerospace, low-orbit satellites, and Beidou System, and how to benefit more people, require the participation of a large number of private enterprises.

For example, low-orbit communication satellites involve too many applications such as material manufacturing and communications services, and all these related industries need to improve. For example, the space station has a very comprehensive pulling effect and it is very worthwhile to follow the national plan to promote industrialization.

Tradewind Think Tank: What is the next commercialized direction for private enterprises in the spaceflight field? How can they form a complementary industrial ecology with the national teams?

Xie Gengxin: Commercial spaceflight is currently in its infancy in China. The most important thing is to find its own position and complement the national teams while responding to the urgent needs of the country.

For example, if a private rocket company wants to make products like the Long March series of rockets, it is definitely not in line with reality, because it neither has the ability to concentrate such a large number of people and resources, nor the long-term accumulation of technology and experience.

So what can private enterprises do? They can make efforts on the professional side to explore whether commercial spaceflight can be made deeper and more detailed in a certain specialty or a certain point. For example, some rocket companies are now trying to use different fuels ratios, focusing on new materials and new fuels, and some private enterprises are making low-cost and lightweight rockets. Therefore, commercial spaceflight and private enterprises must work hard on their professions and focus on the niche markets instead of making large and complete products, because it is difficult.

The second is to complement and support the national teams so that the national spaceflight teams can focus on major tasks. For example, in the field of architecture, the design institute is responsible for the design of the house, the construction company is responsible for the construction, and the production of building materials and the organization of the construction team are all handed over to the outsourcing partner.

Private enterprises can manufacture some dedicated satellites and application-oriented satellites. After the development of satellite is completed, it will immediately serve the citizens or industry. In the production process, private enterprises are more flexible in management. They can use new technologies, such as 3D printing and integrated molding so that they can truly form a complete set of national teams, and can also do more profoundly and farther in certain professional aspects.

Tradewind Think Tank: Musk's Starlink project has made low-orbit communications satellites a hot topic currently. Then how should we view the global development of low-orbit communications satellites?

Xie Gengxin: Space X plans to launch 42,000 satellites. In fact, China is also planning in this regard. Low-orbit satellites are definitely a trend, but how it develops and where it goes still needs to be explored.

Is there a better way to build this kind of low-orbit communication? Low-orbit communications satellites have a limited lifespan in space, so how to clean them in space when they end the service life? This involves a series of problems, but now there seems to be no good solution.

The exploration of space resources requires global community to jointly discuss a more advanced, more feasible, and more scientific plan. Space resources must be very scarce resources. Space resources are not like land resources. For example, China's territory belongs to China only. The Earth rotates on its own and space is variable and three-dimensional. Our laws and our thinking on the development and utilization of space resources have not kept up with our technological development.

Next, we should seriously consider how to develop and utilize space resources effectively. Not only China, but the whole world needs to think about this issue. Development should not be pursued by brute force, but should be pursued in the interest of human development, human well-being, and shared interests.

Jin Xianmin:

How Quantum Technology Ushers in a New Era of Mankind

What kind of changes did the first and second quantum technology revolutions bring to us respectively? Why do we say that quantum communication can achieve unconditional information security and confidentiality? What is the practical significance of developing optical quantum chips and what is the current research progress? Why are the fields such as biomedicine, financial transactions, and aerospace as excellent application scenarios for quantum computing?

Changes Brought About by Quantum Technology

Tradewind Think Tank: What is the origin of quantum technology? Why is quantum mechanics one of the cornerstones of modern physics?

Jin Xianmin: Quantum technology was raised at the beginning of last century. Scientists turned human cognition of the world from "Newtonian mechanics" to "quantum mechanics" by upgrading theories and made quantum technology ushered in vigorous development.

The quantum mechanics theory significantly supports the quantum technology. And before this, we believed that what we see is what we get. Especially with the formation of the "grand unified theory (GUT)" of the four forces: gravity, electromagnetism, strong nuclear forces, and weak nuclear forces, many scientists believed that besides the two "clouds," Ether and Black-body Radiation, physics was certain and that human basically grasped the whole rules of universe.

So what is Black-body Radiation? It is known that any object above absolute zero (about -273.15 °C) will produce radiation, which is a conduction method in which objects use electromagnetic waves to radiate heat energy outward. As the temperature of the hot material rises, electromagnetic waves begin to intensify and produce visible light of color changing to red, yellow and blue in turn. However, the question of how much energy the electromagnetic waves of each color have has always plagued scientists. In order to solve this problem, Planck, the father of quantum mechanics, proposed in 1900 that "The energy of electromagnetic radiation is confined to indivisible packets (quanta)" which is the smallest indivisible unit in a physical quantity through which the energy value of electromagnetic waves can be calculated. Planck's theory opened the curtain of the quantum world.

Soon after, scientists such as Einstein, Heisenberg, Schrödinger, and Bohr gradually improved the theoretical framework of quantum mechanics. In classical physics, Newtonian mechanics can accurately predict the position and velocity of objects. In 1926, Schrödinger proposed to view the quantum world using the wave equation, which refers that all particles do not have a defined position until they are observed, giving the probability of finding a particle at a given point. On this basis, Bohr and Heisenberg proposed the "Copenhagen Interpretation." According to the Copenhagen Interpretation, wave function collapse will appear and particles in the superposition state will be in a defined position when measuring a physical system. In other words, it is the observation of the observer that changes the system.

In China, there is a joke on the Internet that "When things are knotty, turn to quantum mechanics," which is fun but reflects the significance of quantum mechanics theory. It determines the way we perceive the world. And the microscopic phenomena of chemistry and biology still need to be explained by the quantum mechanics. Except for philosophy, which cannot be explained by "quantum", almost all natural sciences are expected to understand the material world in depth through the theory of quantum mechanics. Quantum mechanics has become one of the important foundations of human understanding of the world.

Tradewind Think Tank: What kind of changes did the first and second quantum technology revolutions bring to us respectively?

Jin Xianmin: Around the 1950s, the process of studying the laws of quantum mechanics gave rise to many modern information technologies, and the first quantum technological revolution was born.

The most important contribution of the first quantum revolution is the semiconductor whose conduction capacity lies in between that of conductors and insulators at room temperature. A wall of resistance appears when a semiconductor comes into contact with a metal and electrons can pass through this wall of resistance when they are in motion. In the field of classical mechanics, the phenomenon of an object passing through a wall cannot be explained, but quantum mechanics can be explained by the "tunneling effect." For example, in the routers we daily use, electromagnetic waves emitted by their wireless signals can move through walls.

Without the understanding of semiconductors based on quantum mechanics, we couldn't invent silicon transistors, microchips, micro-computers, and global positioning systems. In a sense, quantum mechanics is the theoretical foundation of modern information technology. The first quantum technological revolution directly gave birth to the booming development of modern information technology.

And how far-reaching is the impact of this quantum technology revolution on our world? Some people have done statistics that 70% of the national economy is now related to quantum technology, which shows the far-reaching impact of quantum technology.

In 1989, quantum technology ushered in a second technological revolution as advances in quantum-related experimental techniques made it possible for humans to actively manipulate and measure individual quantum states. We can strip, manipulate, and detect single photons, but this process is actually very difficult. For example, a 15-watt electric bulb can emit 1,021 single photons per second, all of whose information need to be processed.

As the information carrier of quantum, light wave is a good breakthrough. Human could advance the research of quantum from "passively observing and applying its laws" to "actively regulating and manipulating its state" by studying the characteristics of wave-particle duality, multi-particle superposition and entanglement of light at room temperature, and further master the construction of the entire quantum world of each block. From here, the second quantum revolution bursts.

The second quantum technology revolution has resulted in three core technologies: quantum communication, quantum computing and quantum precision measurement. Quantum communication can realize almost unconditional secure communication, which is an unprecedented change in communication network and information security. Quantum computing breaks through the physical limitations of silicon-based computers and exponentially enhances computing capabilities, greatly changes the IT industry. Quantum measurement is a precision measurement based on microscopic particle systems and their quantum states, which greatly surpasses traditional measurement techniques in terms of accuracy, sensitivity, and stability.

Tradewind Think Tank: Why is it said that quantum communication can achieve unconditional information security and confidentiality?

Jin Xianmin: Quantum communication is a new type of communication method for information transfer, which can achieve unconditional information security and confidentiality through "Quantum Key Distribution (QKD)" + "One Time Pad."

QKD is quantum cryptography, taking advantages of unclonable function and Heisenberg's uncertainty principle of quantum states. The principle of unclonable quantum state means that a single quantum cannot be divided. Once manipulated or measured, the quantum energy state will be destroyed. Therefore, any quantum state cannot be accurately replicated. Heisenberg's uncertainty principle imposes a restriction on the accuracy of simultaneous measurement of position and momentum. The more preciseis, the less accurate will be our momentum measurement and vice-versa.

Based on these, Charles Bennett and Gilles Brassard at IBM proposed the "BB84 Protocol" in 1984. BB84 is a quantum key distribution scheme and the first quantum cryptography protocol. The protocol is provably secure, relying on the quantum property that information gain is only possible at the expense of disturbing the signal if the two states one is trying to distinguish are not orthogonal and an authenticated public classical channel. It is usually explained as a method of securely communicating a private key from one party to another for use in one-time pad encryption.

How was it implemented? It is known that a bit is the basic unit of information in a classical computer, and a quantum bit is the basic unit of quantum information, which only adds to the quantum properties of physical atoms. The data carrier of quantum communication is the quantum bit encoded by polarization. Considering the randomness of quantum, it is possible for receiver to encounter error when transmitting each polarization-encoded random bits composed of 0 and 1. And both parties communicate through classical communication to delete the wrong bits, and the remaining are the correct passwords randomly generated.

However, if an unauthorized third party intercepts the qubits, the final results are often very different. And a comparison of part of the data between the two communicating parties will reveal a higher error rate, which proves that there is an eavesdropper.

In addition, One Time Pad is the only symmetric encryption mechanism that has been proven to be "absolutely secure," but requires the use of a single-use pre-shared key that is no smaller than the message being sent. In this technique, a plaintext is paired with a random secret key (also referred to as a one-time pad). Then, each bit or character of the plaintext is encrypted by combining it with the corresponding bit or character from the pad using modular addition. Quantities of safe keys are required to meet such encryption standard, which makes it more difficult to eavesdrop.

Another type of quantum communication is quantum stealth transmission. Simply put, a single photon is indivisible, and photon A will affect photon B no matter it spins towards left or right because there is an unknown connection between them that even exceeds the speed of light which is called "Quantum Entanglement."

In quantum stealth transmission, the two communicating parties, A and B, who are in two distant places respectively, first share a pair of entangled particles. A distinguishes the particles of the quantum state to be transferred, and then informs B of the result, and B performs the corresponding operation according to the information obtained.

With the "help" of quantum entanglement, the quantum state to be transferred undergoes a "hyperspace transfer" as described in science fiction, in which the original remains at the sender and only the quantum state of the original is transferred.

But at present, quantum stealth transmission is still in the laboratory stage, which is far from practical application.

Tradewind Think Tank: How is the industrialization of quantum communication? What are the challenges in this process?

Jin Xianmin: Quantum communication is one of the earliest technologies in the quantum technology industry. It currently aims at implementing quantum communication network in metropolitan area through optical fiber, inter-city quantum communication through relay, and long-distance quantum communication through satellite relay.

China is developing rapidly in the field of Quantum communication. China launched the world's first quantum scientific experiment satellite "Mozi" in 2016, building the world's first quantum private communication line, "Beijing-Shanghai trunk," in 2017, and successfully distributed satellite-to-earth quantum keys across 4,600 km in 2021. The prototype of the integrated wide-area quantum communication network between heaven and earth was constructed.

Of course, there are still some problems in quantum communication, especially the encoding and transmission methods. The current quantum communication terminal, with its huge physical size and high development cost, can only be applied to financial institutions and government departments, and cannot be popularized. But in fact, quantum communication is very useful for communication security in fields such as the Industrial Internet, smart home, and autonomous driving.

We are also thinking about how we can really achieve low-cost, easy-to-use quantum communication for universal application. And the key is to realize the chip-based quantum communication technology, implanting chips into various small and micro terminal devices, developing low-cost, mobile, safe and reliable products and applications, and then forming local or branch networks complementary to backbone-level quantum communication networks.

Based on this direction, we have already realized the application trials of quantum communication in smart home and the Industrial Internet, which includes image transmission and flight control of UAV. These application scenarios have great demands for communication security, which is a trillion-dollar market. We hope that quantum communication can empower the urban IoT and Industrial Internet and can truly benefit millions of households.

Tradewind Think Tank: Why quantum computing can exponentially enhance computing power?

Jin Xianmin: Quantum computing is the most important contribution of the second quantum revolution, which can significantly improve the computing power and intelligence level of society, penetrate into human's daily life, and have a greater significance for the whole world than quantum communication and measurement.

Its principle can be explained in a popular way. In classical computers, classical bits have only two states—0 and 1, but in quantum computing, qubits are no longer simple 0 and 1, but an expanded two-dimensional space superposition state. One qubit will expand to a two-dimensional space superposition state, and two to four-dimensional. If there are N qubits, the expanded spatial superposition state is RN (R=dimension, N=qubits). If so many particles are fully interconnected, they can form a simultaneous superposition or parallel space, which is called "Hilbert Space," to process information.

Classical computers have always followed Moore's Law and now limited by the ceiling of physics. But the potential of quantum computing is so huge that, according to the RN formula, an extra qubit can exponentially increase the computing power of the computer. This contributes to superiority of quantum that surpasses the calculation speed of classical computers. In calculating RSA, an encryption algorithm, even using the most powerful classical computers currently, it would take more than 3 million years to crack the RSA algorithm as it uses the highest cryptographic level 1024-bit binary cipher, while it only takes a few days to use a quantum computer of the same size.

Such technology innovative direction of quantum computing will lead human history into a new era. But it should be taken progressively to achieve ideal goals. What we need to do is to keep researching theory and technical application, and keep investing substantial funds and talents to verify in various application scenarios.

The Road to Industrialization of Quantum Technology

Tradewind Think Tank: Considering your own experience, could you please tell us the whole process from frontier exploration and research, to technology realization and industrialization practice?

Jin Xianmin: Quantum technology research is a tough but promising road. I started to study quantum technology in 2003 at the University of Science and Technology of China. And topic of my doctoral dissertation is to realize the quantum stealth transmission in free space by building a 16-kilometer quantum channel system from the Badaling Great Wall to Huailai, Hebei Province.

The high density of dust in the horizontal atmosphere significantly attenuates the light, causing severe photon loss when penetrating the horizontal atmosphere. In contrast, there is less dust in vertical atmosphere and weaker impact on light attenuation. The loss of photons penetrating the entire vertical atmosphere is equivalent to that of flying 5–10 km in the horizontal atmosphere.

Therefore, we connected the satellites using photons to penetrate the atmosphere vertically and setting distance of 16 km beyond the prescribed value. As a preliminary project of a quantum communication satellite, it perfectly verified the feasibility of satellite-based quantum communication, and further practically supported the "Beijing-Shanghai quantum communication trunk" and "Mozi" quantum experiment satellite.

In 2008, I continued my postdoctoral work and was responsible for building an experimental platform for quantum entanglement storage based on laser-cooled atoms. Quantum storage is the core of quantum communication and computing. This experiment was extremely hard. Because the speed of light reaches at 300,000 kilometers per second, it required such great patience and practice capability to freeze the entangled photons and put them inside the atoms. It was such a test of skill and patience that I slept in the lab for two months. The complex research experience made me understand the various technical directions and then connect them for flexible application.

From 2010 to 2014, I went to the Department of Physics of Oxford University in the UK for physics research, during which I started to be exposed to quantum computing and optical quantum chips. In 2014, I returned to China to teach at Shanghai Jiao Tong University, while researching quantum communication, computing, and optical quantum chips with the funding of university. In the optical quantum chip, we have achieved a full chain of chip manufacturing, including design, wafer, etching, packaging, testing and tape-out. We gradually industrialized the research results and in 2021 established Turing Quantum, the first optical quantum chip and computer company in China, and soon received Angel Investment of nearly 100 million yuan, gaining the highest fund in angel round among quantum technology companies in China and extending from the basic technology research of quantum chips to the large-scale terminal applications.

Of course, no innovation is a smooth sailing. The innovation of quantum computing is more of a "chicken or egg" problem, and you can't prove the value of this thing until the results are fully available. In 2016, our research team at Shanghai Jiao Tong University, we faced the dilemma of financial constraints and the project was once at a standstill. At the most difficult time, it was the government's support in science and technology innovation that gradually brought our young team out of the predicament.

Tradewind Think Tank: Quantum computing is currently divided into photons, electrons, atoms, and so on, from the carrier. But from the technical view, it is divided into Bose sampling, topology, superconductivity and random circuit sampling. So, what are the characteristics of each form, and what are the future trends and prospects for application implementation?

Jin Xianmin: First of all, there are many carriers for quantum computing. In principle, as long as it can carry discrete energy, the physical system can become a carrier of quantum. From the perspective of industrialization, there are currently three main carriers, namely superconductivity, ion traps and optics.

Relatively, the two carriers, superconductivity and ion trap, have their limitations. Quantum must be placed in a vacuum or ultra-low temperature environment, due to its instability, otherwise the quantum entanglement will gradually be lost over time, resulting in quantum decoherence. But optics is more like a "panacea," as photon is characterized by uncoupling with the environment and can move at room temperature without quantum decoherence. And if superconductors and ion traps try to build a distributed computing system with other systems, the optics is required for intermediate connection.

What's more, measuring the state of a quantum computer will collapse the uncertain state of quantum into a defined position, which requires to design a brand-new algorithm that uses features such as quantum interference and entanglement to obtain the result. Therefore, in the technical view, quantum algorithms for different physical systems have appeared, including Bose sampling, topology, superconductivity, and random circuit sampling.

But there are relatively more algorithms that fits for optical carriers. For example, we can regard Bose sampling as a quantum world of Galton board where a small ball is dropped from above with half possibility to go left or right when passing through a pegboard. And when there are many small balls falling randomly, the distribution of the falling position will be statistical regular.

But in calculating, the mass, angle of incidence, surface curvature and hardness of balls should be considered. And optics has the advantages of being immune to electromagnetic fields, little energy consumption of transmission and conversion, and stable transmission without interfering with each other, almost avoid every factor that plague the ball.

All roads lead to Rome. In the future, different quantum computing systems, such as superconductivity, ion traps and optics, will play their unique advantages in certain aspects to form a hybrid quantum computing system. In this process, optics significantly transform the research into industry and connect different systems as superconductors, ion traps and optics.

At present, optical quantum computing is more favored by market, and PsiQuantum in Silicon Valley, which has the highest funding (over $500 million) in this field, is an optical quantum computing company. During the time at Oxford, team at Bristol University and we were both working on quantum computing on optical quantum chips, and were simultaneously seeking funding from the UK government, which was eventually awarded to our team. Later, the academic leader of the University of Bristol team went to Silicon Valley and established PsiQuantum. I believe that, with similar technology development and China's scientific research base, market potential and policy support, we Turing Quantum will be able to catch up with PsiQuantum.

Tradewind Think Tank: From the perspective of industrialization, what are the advantages of optical quantum chips?

Jin Xianmin: Photons have been studied for many years, and there are many application scenarios that are already relatively mature, and macroscopic optics has been in secondary school textbooks for a long time.

However, in terms of computing applications, there are defects in macro-optics. Firstly, the macro-optical system is very fragile, and results will be different with even a slight change of angle of optical magnifiers and lenses on the optical platform. Secondly, it needs hundreds of optical elements to form a certain optical circuit system, which need a huge optical platform. Thirdly, the construction of the macro-optical calculation system strongly depends on the experience of researchers, just like the relish of a restaurant depends on the craftsmanship of chef, which cannot be standardized and replicated like McDonald's.

Then how to solve these problems? I think it is an inevitable way to integrate photons on the chip by introducing quantum technology to meet the needs of scalability and standardization.

In fact, superconducting and ion trap systems also have chips, but these chips need to be placed in a vacuum or ultra-low temperature, fully isolated from the environment, and the peripheral devices around the chip often weigh tons. However, the integrated optical quantum chip can be placed in a room temperature, which is smaller in size and easier to move. We are trying to make the optical quantum integrated chip the size of an electronic computer motherboard and apply it into various application scenarios to improve the computing power of various application scenarios, with which we are confident.

Tradewind Think Tank: What is the practical significance of developing optical quantum chips and how is it going currently?

Jin Xianmin: First of all, we have to distinguish the concepts of photonic chip and optical quantum chip. Photonic chip uses semiconductor light-emitting technology to generate a continuous laser beam to drive other silicon photonic devices, while optical quantum chip integrates optical quantum line on a substrate to process quantum information.

In addition, the photonic chip is a silicon-based manipulation technology. The use of laser technology can make photons more widely used in computers, but there is still the stranglehold of the lithography machine, which, however, will not appear in optical quantum chips, because the manufacturing of optical quantum chips does not require the improvement of physical limits but simply apply the 14-nanometer general-purpose. Optical quantum chips are not designed to solve the dilemma of integrated circuit chips, but a new computing architecture, a promising development in quantum computing.

However, development of optical quantum chips also has difficulties. The biggest bottleneck is the fineness of the process, of which Turing Quantum is ahead. The first achievement is that Turing Quantum has developed a 3D integrated optical quantum chip, using a femtosecond laser to stereoscopically inscribe on silicon dioxide material. Within a few hundred femtoseconds, the laser can modify the material inside, and create any shape of 3D structure and photonic circuit. The second achievement is that, Turing Quantum has developed thin-film lithium niobate optical quantum chip, which is characterized by a wide spectral range, fast modulation, high integration, low loss and other advantages and is expected to replace silicon photonics and optical modem.

It is with these two photonic chips that we are capable to control the entire spectrum of optical modulation, achieving both 3D integration and high-speed and low-loss modulation, thus promoting the application and development of optical quantum computing, optical artificial intelligence processors, optical communication and interconnection.

It costs us a lot to build a fully integrated optical quantum computing system and break through the research links of optical source, processing and detection, covering the entire chain of optical quantum chip such as design, packaging and tape-out. But why not adopt the mode of division of labor and cooperation, like electronic chips, and outsource the links such as tape-out? Because at present, the global development of optical quantum chips is still in the early stage of continuous iteration, and is not as mature as electronic chips who has a very clear industry chain of labor division in various parts. So it is more conducive to master the core technology and continuously upgrade and iterate to control the whole industry chain of optical quantum chip.

The Advantages of Quantum Computing

Tradewind Think Tank: The number of manipulated qubits is one of the evaluation criteria for quantum computing. But how will this affect the future upgrade of quantum computing?

Jin Xianmin: Currently, Google claims to have achieved quantum supremacy with 53 qubits, while future practical quantum computing will require manipulating dozens or even hundreds of qubits in entangled state at the same time. In quantum computing, the number of qubits that humans can manipulate determines the capability of quantum information processing. However, I believe that there are inaccuracies in manipulating the number of qubits.

The first reason is low of the error tolerance in manipulating the number of qubits. Since qubits can be any combination of 1 and 0, a tiny error in quantum computing or any spurious signals in coupling to the physical system will eventually mistake the calculation. According to statistics in 2018, the error rate of qubit manipulation exceeds several percent on systems with five or more qubits.

The second reason is difficult to control the number of qubits. Each qubit currently has to be connected to several wires to regulate and read. And the pulse waveform generated by the manipulated qubit has to be extremely precise, and the measurement accuracy of the microwave tank must be as accurate as that of a single photon when reading the qubit. In addition, considering the decoherence of qubits, only in a sufficiently long coherence time can more quantum circuits and more complex quantum algorithms could be completed in limited "lifetime" of the qubit. And many teams are currently trying to improve the coherence time with new structures or materials, but it is still a hard nut to crack.

Therefore, a truly rational way needs to be found on how to manipulate the number of qubits. In quantum coding, it does not necessarily have to be encoded into qubits, but can be multi-dimensional quantum units of information, expanding the scale of space. For example, optical quantum computing, according to the aforementioned RN formula, can be encoded into dozens or even hundreds of dimensions. Enlarging the dimensionality of qubits can be a measure of the advanced degree of quantum computing.

Tradewind Think Tank: How will quantum computing empower big data and artificial intelligence?

Jin Xianmin: In big data, quantum computing can achieve more accurate and larger dynamic range of sorting through parallel computing. In 2020, our team at the School of Physics and Astronomy of Shanghai Jiao Tong University has demonstrated the research results of large-scale quantum web page sorting.

Quantum web sorting is a quantum algorithm based on quantum random walk theory. Compared with the web sorting of search engines such as Google, Baidu and Bing, the algorithm of quantum web sorting has more advantages, such as being able to improve the accuracy of sorting results, reduce the congestion of sorting results from different nodes, and distinguish between primary and secondary nodes in the network.

In artificial intelligence, the transformation matrix processed by the quantum computing can be applied to the optical neural network of artificial intelligence to form artificial intelligence judgment with low-power and low-latency. An integrated circuit chip works with many logical operations to finally form an AI judgment, while inside the optical quantum computing, once the optical neural network training is completed, AI judgment can be achieved only with an easy movement of light from the input to the output. Since this is a static algorithm without so many logical judgments, it is fast and of low consumption, and highly applicable in some edge terminals of artificial intelligence. In Cooperative Vehicle Infrastructure System (CVIS), for example, it can achieve rapid decision-making in less than 100 milliseconds, which is several orders of magnitude higher than human response time.

In addition, quantum algorithms have advantages in exploring optimal solutions. One problem with the current slow development of robots is that they learn too slowly and require a lot of data training to find an optimal solution. If a robot wants to acquire "intelligence" quickly, it can surpass humans in determining optimal solutions within just a few months by combining quantum algorithms and artificial intelligence and quickly learning in human society. And in the next 5-10 years, I believe there will be a great prosperous industry in combining quantum computing and optical artificial intelligence.

Tradewind Think Tank: Why do you say that biomedicine, financial transactions, and aerospace are excellent application scenarios for quantum computing?

Jin Xianmin: In biomedicine, it will cost a lot both in time and finance and is full of blindness and contingency to develop a new drug, with a low success rate of clinical research. But with the fast, parallel data processing capabilities of quantum computers, it is possible to accurately assess the interactions between molecules, proteins and chemicals and check whether a drug will improve or cure certain diseases. In the development, screening and optimization of new drug, quantum computing can exponentially increase the efficiency compared to classical computing.

In the context of financial transactions, given the various correlations of thousands of assets, quantum computing can help with portfolio optimization and identify key fraud indicators more effectively. This quantum optimization is now being successfully applied to solve actual financial problems, including optimal trade routes, arbitrage opportunities, and feature selection in credit scoring, where there are variables in the data that classical computers cannot solve quickly and efficiently.

In the aerospace field, a U.S. company, D-Wave, the world's

first commercial quantum computing company, started a deep cooperation with NASA in 2013. In order to cope with the complex and huge interstellar orbit data that traditional computers cannot handle, D-Wave uses quantum annealing algorithm to slow down its collapse and retain longer quantum superposition states for quantum computing, so as to process complex and huge interstellar orbit data.

Whether it is bio-medicine, financial transactions, or aerospace, the great computing power is a common demand. Because classical computing cannot quickly and accurately calculate when facing massive, variable, complex data, while quantum computing can perfectly handle to meet the high-dimensional computing needs.

Therefore, quantum computing can demonstrate more superiority in this sort of problems for a long time in the future, as well as positively affect the national economy to which this sort of problems is linked. In particular, quantum technology can improve the accuracy of financial high-frequency transactions, or exponentially accelerate the speed of search engines, and speed up the development of drugs. Quantum computing will improve all aspects of human life in orders of magnitude, and will be popularized from point to point in all industries. Just as the electronic computer that could only encode when it first came out, but the future of quantum computing will definitely reshape human civilization as what electronic computer did.

(According to the online data, the Tradewind Think Tank)

The Development History of Quantum Computers

Time	Project	Contributions
2016	IBM	The first online quantum computer platform worldwide, equipped with a 5-bit quantum processor.
2018	Origin Quantum	The world's most powerful 64-bit quantum virtual machine then, refreshing the world record of classical computers to simulate quantum computers at that time.
2019	IBM	The world's first commercial quantum computer, IBM Q System One, which is the first commercially available quantum processor.
2019	Google	The 53-bit superconducting quantum, taking about 200 seconds to complete the task that would take a classic computer 10,000 years to complete, achieving "quantum supremacy."
October, 2020	Origin Quantum	Officially open the self-developed superconducting quantum computing cloud platform to the public worldwide, which is based on the superconducting quantum computer, Wuyuan, independently developed by Origin Quantum and equipped with the 6-bit superconducting quantum processor, Kuafu KFC6-130.
December, 2020	Team Leaded by Pan Janwei and Lu Zhaoyang of University of Science and Technology of China	Successfully constructed the 76 photon quantum computers, Jiu Zhang, which can achieve the 600 million years of calculations of the world's fastest supercomputer, Fu Yue, in 200 seconds.